MEI STRUCTURED MATHE

SECOND EDITION

Decision and Discrete Mathematics 1

Chris Compton
Keith Parramore
Geoff Rigby
Joan Stephens

Series Editor: Roger Porkess

Hodder & Stoughton

A MEMBER OF THE HODDER HEADLINE GROUP

Acknowledgements

OCR, AQA and Edexcel accept no responsibility whatsoever for the accuracy or method of working in the answers given.

AQA (AEB) examination questions are reproduced by permission of the Assessment and Qualifications Alliance.

Orders: please contact Bookpoint Ltd, 130 Milton Park, Abingdon, Oxon OX14 4SB. Telephone: (44) 01235 827720, Fax: (44) 01235 400454. Lines are open from 9.00–6.00, Monday to Saturday, with a 24 hour message answering service. Email address: orders@bookpoint.co.uk

British Library Cataloguing in Publication Data
A catalogue record for this title is available from the The British Library

ISBN 0 340 771895

First published 1992
Second edition 2000
Impression number   10 9 8 7 6 5 4
Year   2005 2004 2003 2002

Copyright © 1992, 2000 Chris Compton, Keith Parramore, Geoff Rigby, Joan Stephens

Typeset by Pantek Arts Ltd, Maidstone, Kent.
Printed in Great Britain for Hodder & Stoughton Educational, a division of Hodder Headline Plc, 338 Euston Road, London NW1 3BH by J.W. Arrowsmiths Ltd, Bristol.

MEI Structured Mathematics

Mathematics is not only a beautiful and exciting subject in its own right but also one that underpins many other branches of learning. It is consequently fundamental to the success of a modern economy.

MEI Structured Mathematics is designed to increase substantially the number of people taking the subject post-GCSE, by making it accessible, interesting and relevant to a wide range of students.

It is a credit accumulation scheme based on 45 hour modules which may be taken individually or aggregated to give Advanced Subsidiary (AS) and Advanced GCE (A Level) qualifications in Mathematics, Further Mathematics and related subjects (like Statistics). The modules may also be used to obtain credit towards other types of qualification.

The course is examined by OCR (previously the Oxford and Cambridge Schools Examination Board) with examinations held in January and June each year.

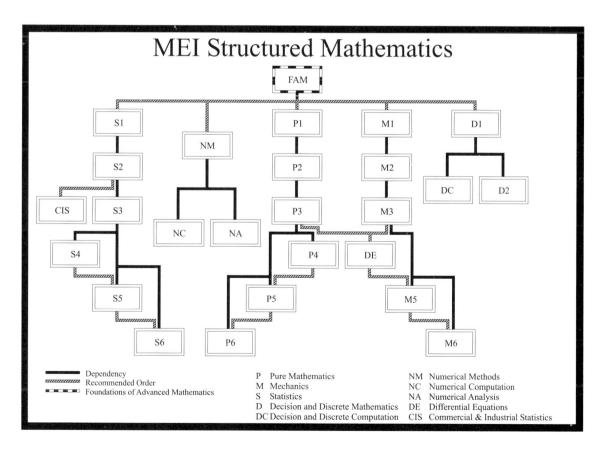

This is one of the series of books written to support the course. Its position within the whole scheme can be seen in the diagram above.

Mathematics in Education and Industry is a curriculum development body which aims to promote the links between Education and Industry in Mathematics at secondary level, and to produce relevant examination and teaching syllabuses and support material. Since its foundation in the 1960s, MEI has provided syllabuses for GCSE (or O Level), Additional Mathematics and A Level.

For more information about MEI Structured Mathematics or other syllabuses and materials, write to MEI Office, Albion House, Market Place, Westbury, Wiltshire, BA13 3DE.

Introduction

This is the first in a set of three books written to support the Decision and Discrete Mathematics modules in MEI Structured Mathematics. It is supported by additional on-line materials available through the MEI office: Office@mei.org.uk.

Together the books cover the basic work in this branch of mathematics and so are also suitable for other courses in this strand of AS and A Level Mathematics. Throughout the emphasis is on understanding and interpretation.

Algorithms form the basis of this branch of mathematics and so they are the subject of the opening chapter of this book. You meet a number of questions.

- What are algorithms?

- How do you communicate algorithms?

- How do you measure the efficiency of an algorithm?

Algorithms form an extensive area of study in their own right, and so their treatment is essentially an introduction. The same is also true of graphs, networks and linear programming, which are continued in later books in the series.

This book also includes a chapter on critical path analysis and this topic is particularly suitable for meeting the coursework requirements of this module. Detailed advice about coursework is available from the MEI office.

This is the second edition of this book. Some parts of it are little changed but others (such as the chapter on linear programming) are completely new, in line with the changed specification requirements. I would like to thank all those who have been involved in writing it, and also those who have offered advice on it, particularly Ian Bloomfield and Charlie Stripp.

Roger Porkess
Series Editor

Contents

Preface

Why 'Decision and Discrete' mathematics? What do the terms mean?

Discrete mathematics contrasts with continuous mathematics, which is concerned with the real number system. Here limits exist and the techniques of calculus are available. Discrete mathematics relates to other number systems such as the integers and the rationals (fractions). In continuous mathematics, quantities are measured or weighed. In discrete mathematics, quantities are determined by *counting*. Thus you would refer to an *amount* of, say, liquid (continuous), but to a *number* of people (discrete).

Decision mathematics is, broadly, the application of mathematical modelling to solve real world problems, often arising from managing commercial and industrial concerns. To use the power of mathematics to solve problems you first need to capture the essence of the real world problem in mathematical form. This move from the real world into the world of mathematics is known as mathematical modelling. It requires simplifying assumptions – so that the mathematical problem which is extracted is tractable.

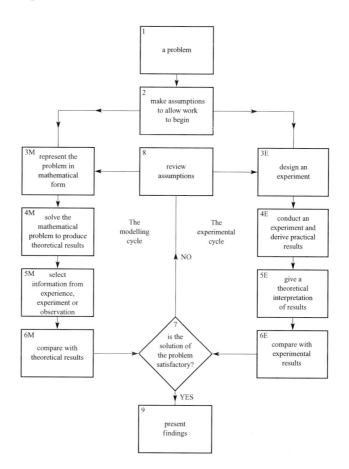

Having extracted a mathematical problem the next stage is to use mathematical techniques to solve the problem.

Management problems usually require decisions to be made, and more often than not they lead to mathematical problems which are discrete in nature. Thus the techniques are not usually calculus based, but involve algorithmic approaches – efficient ways of finding the best out of many possible decisions.

But it is not enough to solve the mathematical model. The solution must be interpreted back into the real world to see if it points to a solution to the original real world problem. At first attempt it probably will not – the essential simplifying assumptions may result in the problem not having been completely encapsulated by the mathematics. If this is the case then a second loop around the modelling cycle (iteration) will be needed. Here the assumptions are reviewed and modified, and the consequences followed. A third or fourth cycle may be needed, as many as are required until an acceptable solution is achieved.

Finally a report has to be completed. It is of no benefit to have worked through the process and arrived at a satisfactory conclusion if the manager who has responsibility for making the decision cannot be convinced of the reliability of your work. So there is a need to be able to communicate complex ideas clearly, succinctly, yet thoroughly.

Algorithms

Monday's child is fair of face,
Tuesday's child is full of grace,
Wednesday's child is full of woe,
Thursday's child has far to go,
Friday's child is loving and giving,
Saturday's child works hard for its living,
And a child that is born on the Sabbath day,
Is fair and wise and good and gay.

Do you know on which day of the week you were born? If not, then you could use Zeller's algorithm to work it out from your date of birth. An algorithm is simply a sequence of precise instructions to solve a problem.

Zeller's algorithm	Example: 15 May 1991
Let day number $= D$ month number $= M$ and year $= Y$.	$D = 15$ $M = 5$ $Y = 1991$
If M is 1 or 2 add 12 to M and subtract 1 from Y	
Let C be the first two digits of Y and Y' be the last two digits of Y.	$C = 19$ $Y' = 91$
Add together the integer parts of $(2.6M - 5.39)$, $(Y'/4)$ and $(C/4)$, then add on D and Y' and subtract 2C. (Integer part of 2.3 is 2, of 6.7 is 6, i.e. the whole number part, but note that integer part of -1.7 is -2 and -3.1 is -4 etc.)	$7 + 22 + 4 + 15 + 91 - 38 = 101$
Find the remainder when this quantity is divided by 7.	3
If the remainder is 0 the day was a Sunday, if it is 1 a Monday etc.	$\therefore$ 15 May 1991 was a Wednesday

❷ Try Zeller's algorithm for yourself using your own date of birth.

Many algorithms are rather tedious to work through by hand, and can be written in such a way that a computer can carry out the task for us. Computer programming languages differ, but whatever language is available it is helpful first to express an algorithm in pseudo code – a cross between English and computer syntax.

Zeller's algorithm could be written in pseudo code as:

Step 1   Let D be Day number (e.g. 4)
          Let M be Month number (e.g. 11)
          Let Y be Year number (e.g. 1985)

Step 2   If $M < 3$ then $M := M + 12$ and $Y := Y - 1$

Step 3   Let $C = \text{INT}(Y/100)$
          Let $Y' = Y - (100 \times C)$

Step 4   Let $S = \text{INT}(2.6 \times M - 5.39) + \text{INT}(Y'/4) + \text{INT}(C/4) + D + Y' - (2 \times C)$

Step 5   Let $\text{Day} = S - (7 \times \text{INT}(S/7))$

If you have done some programming before you can see that this will just produce the day number, not its name. You might like to add some statements to get the output as a day of the week rather than a number.

You will have been using algorithms since you first went to school. The algorithms that you were taught for long multiplication and division would result in working like this:

$$
\begin{array}{r}
163 \\
\times 24 \\
\hline
3260 \\
652 \\
\hline
3912 \\
\end{array}
\qquad
\begin{array}{r}
163 \\
24\,)\overline{\,3912} \\
24 \\
\hline
151 \\
144 \\
\hline
72 \\
72 \\
\hline
0 \\
\end{array}
$$

The multiplication algorithm involves such instructions as:
write down a 0;
multiply the top row by the tens digit; etc.
It is a challenging exercise to write down precise instructions for these two algorithms: you have become so used to using them that you do so without really thinking.

The word 'algorithm' has become more commonplace since the development of the computer. A computer program is simply an algorithm written in such a way that a machine can carry it out. Our interest in this course is in mathematical algorithms, but cookery recipes, knitting patterns and instructions for setting the video to record your favourite TV programme could all be said to be algorithms. People, like machines, can work through algorithms automatically and can be led to the solution of a problem without needing to understand the process. Try writing a set of instructions to enable an eleven-year-old, who knows nothing of algebra, to solve a pair of simultaneous equations.

Communicating an algorithm

How do we communicate our algorithms? In Zeller's algorithm we used ordinary language and pseudo code. The form of communication you choose depends on to whom (child, scientist, etc.) or to what (computer, programmable calculator) you are trying to convey the algorithm. In all cases you must consider both the language you use and the nature of the steps into which you break down the process.

Let us take as an example an algorithm to find the real roots of the quadratic equation

$$ax^2 + bx + c = 0 \qquad \text{(assume } a \neq 0\text{)}$$

using the formula

$$x = \frac{-b \pm \sqrt{b^2 - 4ac}}{2a}.$$

Using *pseudo code* you could write:

Step 1   Let $d = b^2 - 4ac$.

Step 2   If $d < 0$ print 'no real solutions' and go to Step 4.

Step 3   Let $X_1 = \dfrac{-b + \sqrt{d}}{2a}$.

               Let $X_2 = \dfrac{-b - \sqrt{d}}{2a}$.

Step 4   Stop.

❓ How would you program this on your calculator (assuming it is programmable)?

Another common method of communicating the steps in an algorithm is to use a *flowchart*. A flowchart is a diagrammatic representation of the sequence of steps in an algorithm. It can be particularly helpful in showing the structure where there are large numbers of conditional statements involved.

For our quadratic equation example the flowchart would look like the one in figure 1.1.

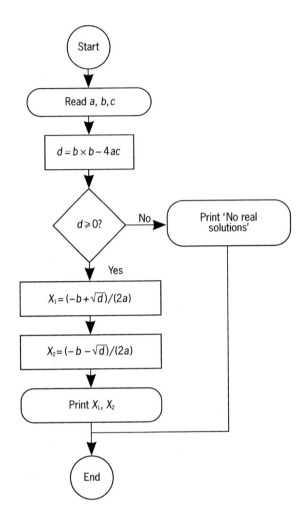

Figure 1.1

Below are two algorithms, one expressed in pseudo code and the other as a flowchart. Consider representing them differently, defining the target user for whom (or which) each representation would be suitable.

RUSSIAN PEASANTS' ALGORITHM FOR LONG MULTIPLICATION

Step 1   Write down the two numbers to be multiplied side by side.

Step 2   Beneath the left number write down double that number. Beneath the right number write down half of that number, ignoring any remainder.

Step 3   Repeat Step 2 until the right number is 1.

Step 4   Delete those rows where the number in the right column is even. Add up the remaining numbers in the left column. This is the result of multiplying the two original numbers.

Here is an example.

24	163
48	81
~~96~~	~~40~~
~~192~~	~~20~~
~~384~~	~~10~~
768	5
~~1536~~	~~2~~
3072	1
3912	

EUCLID'S METHOD FOR FINDING THE HIGHEST COMMON FACTOR (H.C.F.) OF TWO INTEGERS *x* AND *y*

(In other words, the largest integer that will divide into both x and y.) The method is described in the flowchart in figure 1.2.

If you try the flowchart with $x = 24$ and $y = 32$, you should get the output 8, which is the H.C.F. of 24 and 32.

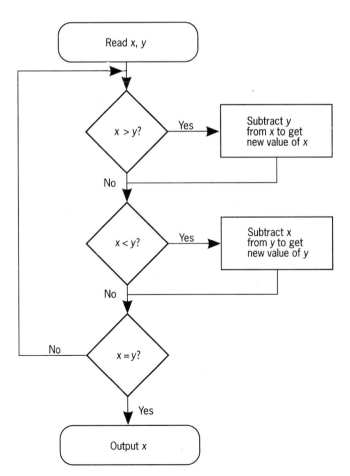

Figure 1.2

Algorithmic complexity

Most problems can be solved using a variety of algorithms, some of which might be more *efficient* than others. By efficient we usually mean using fewer operations, but there might be other considerations too. For example, if the algorithm needs to be run on a computer, you will want to know how much storage capacity it needs.

As a simple example of improving efficiency, consider again the algorithm to find the real roots of the quadratic equation

$$ax^2 + bx + c = 0$$

using the formula

$$x = \frac{-b \pm \sqrt{b^2 - 4ac}}{2a}.$$

It is a good idea to calculate the value of $b^2 - 4ac$ as a first step, because the sign of that value has to be checked to see whether it is worth going on with the calculation. If it is, the value $b^2 - 4ac$ has to be used twice in the calculation.

If an algorithm requires the evaluation of a quadratic expression, the way in which the expression is written can make a difference to efficiency.

For example $3x^2 + 2x + 9$ ①

can be written as $(3x + 2)x + 9$.   ②

Check for yourself that the two expressions are equivalent.

When $x = 5$, evaluation with a calculator requires the following entries:

Using ① $3\boxed{\times}5\boxed{\times}5\boxed{+}2\boxed{\times}5\boxed{+}9\boxed{=}$ 3 multiplications and 2 additions.

Using ② $3\boxed{\times}5\boxed{+}2\boxed{=}\boxed{\times}5\boxed{+}9\boxed{=}$ 2 multiplications and 2 additions.

Now let us investigate $3x^3 + 2x^2 + 3x + 9$ when $x = 5$.

Check that the expression $((3x + 2)x + 3)x + 9$ is equivalent to $3x^3 + 2x^2 + 3x + 9$.

You need to compare the number of multiplications and additions required to evaluate the expressions when $x = 5$.

Using the original expression you have

$3\boxed{\times}5\boxed{\times}5\boxed{\times}5\boxed{+}2\boxed{\times}5\boxed{\times}5\boxed{+}3\boxed{\times}5\boxed{+}9\boxed{=}$ 6 multiplications and 3 additions.

Using the second expression (known as *nested form*) you have

$3\boxed{\times}5\boxed{+}2\boxed{=}\boxed{\times}5\boxed{+}3\boxed{=}\boxed{\times}5\boxed{+}9\boxed{=}$ 3 multiplications and 3 additions.

❓ Generalise this to see how many multiplications and how many additions are needed to evaluate $\quad a_n x^n + a_{n-1} x^{n-1} + \ldots + a_1 x + a_0$

(where the a's are given numbers) for a given value of x, using each approach.

You should find that using the straightforward first approach needs $\frac{n(n+1)}{2}$ multiplications and n additions. Using nested form is much more efficient, needing only n multiplications and n additions.

Of course, if any of the a's happened to be 0, then some work would be saved, but the focus is on the worst case situation when analysing how an algorithm performs. Because there is an n^2 term in the number of multiplications required using the first approach, the complexity of that algorithm is *quadratic*.

Note

You would say that the algorithm has *quadratic complexity*.

To appreciate the effect of this think about how many multiplications would be required to evaluate a degree 100 polynomial. Using the $\frac{n(n+1)}{2}$ formula it is $\frac{100 \times 101}{2} = 5050$.

Now double the size of the problem and consider a degree 200 polynomial. The number of multiplications using the first approach is $\frac{200 \times 201}{2} = 20100$. Doubling the size of the problem has resulted in almost four times the effort.

The second approach, nested form, has *linear complexity*. For such an algorithm doubling the size of the problem doubles the effort involved.

Algorithms are needed to solve large, often very large-scale problems. The analysis of algorithmic complexity is important in such situations. During this course, you will be expected to consider the complexity of some of the algorithms you meet.

Heuristic algorithms

The bin-packing problem

You are faced with the problem of fitting a number of boxes of the same width and depth but different heights into a rack. The rack is the same depth as the boxes. It is divided into slots of the same width and of a fixed height as shown in figure 1.3.

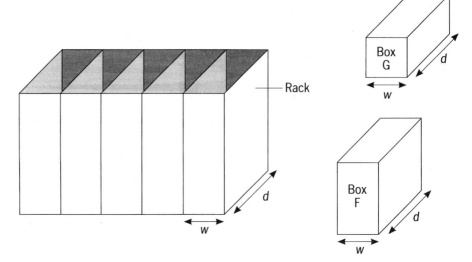

Figure 1.3

There are 11 boxes, A to K, with heights (in dm) as follows.

A	B	C	D	E	F	G	H	I	J	K
8	7	4	9	6	9	5	5	6	7	8

The rack is 15 dm high and you are to stack the boxes one on top of the other using as few slots as possible.

Similar problems might be cutting lengths of wood from standard length planks, or fitting vehicles into lanes on a car ferry. In each case, you are trying to make best possible use of the space available and avoid waste in the form of unused space in the racks, offcuts of wood and unfilled lanes on the ferry.

You can solve all of these problems by a process of trial and error and you will probably have no trouble coming up with the best answer (usually referred to as the *optimal solution*). However, solving a large problem of this type is not so easy. It is worth looking for a more systematic approach to finding the best answer, particularly one that could be carried out by a computer. Later you will be looking at a number of other problems that require development of computer algorithms if large amounts of data are to be handled.

For bin-packing problems the only known algorithms that are guaranteed to find the best solutions are all equivalent to complete enumeration, i.e. looking at all of the possibilities. For large problems this is prohibitively expensive: for m objects in n bins there are n^m possibilities to consider. However, there are many algorithms which will usually find a good, if not the best, solution: these are called *heuristic algorithms*. This raises another issue in our discussion of what constitutes an efficient algorithm. Since you cannot guarantee getting the best solution, if you are choosing from a range of algorithms you would want one that most consistently gave better solutions than the others. The consistency with which an algorithm gives a good solution is therefore another factor in its efficiency.

1 Try to find the best solution to the bin-packing problem above by trial and error. What is the minimum number of slots that is required? This would be referred to as a *lower bound* for the solution.

2 Did you adopt any specific strategy to solve the problem, and if so could you turn it into an algorithm? Try to write your method down such that an eleven-year-old could follow it.

3 How much wasted space is there in your solution?

INVESTIGATIONS

Here are three heuristic algorithms for you to try out. You will need to apply the algorithms to a variety of examples to be able to form a judgement about their relative efficiency, so use them first on the bin-packing problem then on the plumbing, ferry loading and disc storage problems given below. (Answers are provided.)

1 FULL-BIN ALGORITHM

Look for combinations of boxes to fill bins. Pack these boxes. For the remainder, place the next box to be packed in the first available slot that can take that box.

2 FIRST-FIT ALGORITHM

Taking the boxes in the order listed, place the next box to be packed in the first available slot that can take that box.

3 FIRST-FIT DECREASING ALGORITHM

(i) Reorder the boxes in order of decreasing size.
(ii) Apply the first-fit algorithm to this reordered list.

THE PLUMBING PROBLEM

A plumber is using lengths of pipe 12 feet long and wishes to cut the following lengths.

Length (ft)	Number
2	2
3	4
4	3
6	1
7	2

What is the best way of achieving this so that he wastes as little pipe as possible?

THE FERRY LOADING PROBLEM

A small car-ferry has four lanes, each 20 m long. The following vehicles are waiting to be loaded.

Petrol tanker	14 m	Car and trailer	8 m
Car	4 m	Car	3 m
Range Rover	5 m	Coach	12 m
Car	4 m	Lorry	11 m
Car	3 m	Car	4 m
Van	4 m	Lorry	10 m

Can all these vehicles be taken on one trip?

THE DISC STORAGE PROBLEM

A software manufacturer wants to fit the following computer programs on to four 400 kB discs.

Program	A	B	C	D	E	F	G	H	I	J
Size (kB)	100	80	60	65	110	25	50	60	90	140

Program	K	L	M	N	O	P	Q	R
Size (kB)	75	120	75	100	70	200	120	40

Can this be done?

Which algorithm have you found to be the most efficient?
Which one do you think would be easiest to automate, and why?

You have probably been drawing diagrams or possibly working with cut-out shapes. The following is a computational procedure that you could use for the first-fit or first-fit decreasing algorithms.

Define a list of numbers, P, for the heights of the packages (ordered if necessary).

For the bin-packing example, P = {8, 7, 4, 9, 6, 9, 5, 5, 6, 7, 8}.

Define a second set of numbers, B, for the space remaining in the bins. At the very worst this list will need to be as long as the list of packages. For this example B = {15, 15, 15, 15, 15, 15, ...} initially.

Now follow the steps described below:

Step 1	Take the first entry in P.	
Step 2	Is it less than or equal to the first entry in B?	Yes → Step 4 No → Step 3
Step 3	Go to next B entry. Is it less than or equal to this entry in B?	Yes → Step 4 No → Step 3
Step 4	Reduce the B entry by this amount.	

Step 5   Any more entries in P?   Yes → Take next
                                                     entry in P, go to Step 2
                                                     No → Stop.

The result of applying this algorithm is shown below.

P = {8̌, 7̌, 4̌, 9̌, 6̌, 9̌, 5̌, 5̌, 6̌, 7̌, 8̌}
B = {1̶5̶, 1̶5̶, 1̶5̶, 1̶5̶, 1̶5̶, 1̶5̶, ...}
       7̶ 1̶1̶ 9̶ 1̶0̶ 9̶   7
       0   2   0   5   2

It is easy to see how many bins are used, how much free space there is in each and
what packages are in each bin. It is a fairly easy task from here to produce a
computer program to automate the process.

1 Construct a flowchart to clarify whether a married woman has reached an age
   at which she is eligible for a pension when the regulations are as follows.

   The earliest date on which a woman can draw a retirement pension is 60.
   On her own national insurance she can get a pension if she has already retired from
   regular employment. If not, she must wait until she retires or reaches the age of 65.
   At the age of 65 the pension is paid irrespective of retirement. On her husband's
   national insurance, however, she cannot get a pension until he has reached the
   age of 65 and retired from regular employment, or until he is 70 if he does not
   retire before that age.

2 Given below is a table used in an algorithm for converting Roman to decimal
   numbers.

	M	D	C	L	X	V	I
1	1000/2	500/3	100/9	50/5	10/10	5/7	1/11
2	1000/2	500/3	100/9	50/5	10/10	5/7	1/11
3			100/9	50/5	10/10	5/7	1/11
4			100/4	50/5	10/10	5/7	1/11
5				50/6	10/10	5/7	1/11
6					10/6	5/7	1/11
7						5/8	1/11
8							1/8
9	800/5	300/5	100/4	50/6	10/10	5/8	1/11
10			80/7	30/7	10/6	5/8	1/11
11					8/0	3/0	1/8

Take the Roman CIX as an example.

Always start by looking at row 1. Look at the row 1 entry in the column headed C (the first symbol in this Roman number) to find 100/9. This means add 100 into a running total and move to row 9.

Take the second symbol I and look down column I to row 9. You find 1/11, so add 1 to the running total to give 101 and move to row 11. Finally look down column X to row 11 where you find 8/0, so add 8 to the running total to give 109. Since this was the last symbol in the number, we have finished and CIX = 109. (If you end up in a blank square you have made an error.)

Consider writing the algorithm in various ways and use it to carry out some conversions from Roman to decimal numbers. Does the algorithm have any limitations? Can you write an algorithm to convert decimal to Roman numbers?

3 Computer programming languages contain statements that perform repetitions. In Basic the 'for... next' statement works as follows:

```
FOR I = 1 TO 3          FOR I = 1 TO 2
PRINT I                 FOR J = 1 TO 2
NEXT I                  PRINT I, J
                        NEXT J
                        NEXT I
```

The resulting printouts will be as follows.

```
1                       1 1
2                       1 2
3                       2 1
                        2 2
```

In a certain town the bus tickets are numbered 0000 to 9999, and children collect those whose digits add up to 21. How many will there be in a sequence of tickets from 0000 to 9999?

On the next page are two algorithms to solve the problem, written in Basic. Show that each achieves the desired result and compare their efficiency by counting the number of additions/subtractions and comparisons. (**Note:** It is not essential that this question be tackled on a computer.)

```
TOTAL = 0                      TOTAL = 0: SUM = 0
FOR I = 0 TO 9                 FOR I = 0 TO 9
FOR J = 0 TO 9                 FOR J = 0 TO 9
FOR K = 0 TO 9                 FOR K = 0 TO 9
FOR L = 0 TO 9                 FOR L = 0 TO 9
SUM = I + J + K + L            IF SUM = 21 THEN
IF SUM = 21 THEN                  TOTAL = TOTAL + 1
   TOTAL = TOTAL + 1           SUM = SUM + 1
NEXT L                         NEXT L
NEXT K                         SUM = SUM – 9
NEXT J                         NEXT K
PRINT TOTAL                    SUM = SUM – 9
                               NEXT J
                               SUM = SUM – 9
                               NEXT I
                               PRINT TOTAL
```

INVESTIGATIONS

1 Use Zeller's algorithm to show that the 13th of the month is marginally more likely to be a Friday than any other day of the week. The days of the week sequence repeats itself over a 400-year cycle (century years are only leap years if divisible by 400), so consider the 13th of every month from, say, January 1901 to December 2300. Produce a frequency table for the number of occurrences of each day of the week.

2 All books have an ISBN (International Standard Book Number) of 10 digits. The last is a check digit produced by performing a calculation on the other nine. Its purpose is to reduce the likelihood of copying down a wrong number. The check digit is found by multiplying the first digit by 1, the second by 2, the third by 3 and so on up to the ninth, and adding the results, then finding the remainder when this sum is divided by 11. If the remainder is 0 to 9 this digit is the check digit and if the remainder is 10 an X is used as the check digit.

   For example, 019281022 gives

   $$0 \times 1 + 1 \times 2 + 9 \times 3 + 2 \times 4 + 8 \times 5 + 1 \times 6 + 0 \times 7 + 2 \times 8 + 2 \times 9$$
   sum = 117
   remainder when divided by 11 is 7.
   So the full 10-digit number is 0 19 281 022 7.

   Write a computer program to perform this algorithm and check it on some ISBNs.

3 Write an algorithm to find all the divisors of an integer (e.g. the divisors of 24 are 1, 2, 3, 4, 6, 8, 12, 24).

A 'perfect' number is one that is equal to the sum of all its divisors that are less than itself.

The first perfect number is 6 since $6 = 1 + 2 + 3$.

Write an algorithm to find perfect numbers and use it to find the next two perfect numbers. (One if you are working manually! Numerical answer provided.)

4 The knapsack problem

A hiker going on a camping trip can carry a weight of up to 45 pounds in his knapsack. There are five items that he had hoped to take but to take them all would exceed his weight allowance. He has assigned a value to each item. The items and their weights and values are as follows:

Item	Weight	Value
1	13	7
2	12	9
3	15	30
4	16	16
5	18	27

What items should he pack so that their total value is a maximum, subject to the weight restriction? (Answer provided.)

Solve the problem by trial and error and also suggest and test heuristic algorithms that would be useful for larger problems of this type.

5 Figure 1.4 shows a diagram of the Hampton Court Maze. Write an algorithm to assist people in finding their way to the centre. Your algorithm should be general, so try it out on other mazes.

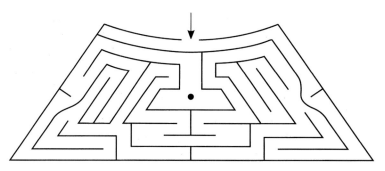

Figure 1.4

6 You may have come across Pascal's triangle

```
                1
            1       1
        1       2       1
    1       3       3       1
                etc.
```

For convenience, number the rows 1, 2, etc. and label the entries in each row as the 1st, 2nd, etc. Thus the third row is 1 2 1 and its 2nd entry is 2.

Here are three ways in which the numbers in the triangle can be produced.

(i) Each new entry is the sum of the two numbers immediately above it.

$$1 \qquad 2$$
$$3$$

(ii) The rth entry in row n is given by the formula

$$\frac{(n-1)!}{(r-1)!(n-r)!}$$

e.g. row 3, 2nd entry $= \dfrac{2!}{1!1!} = 2.$

(iii) Row n has

1st entry always 1

2nd entry $= \dfrac{n-1}{1} \times$ 1st entry

3rd entry $= \dfrac{n-2}{2} \times$ 2nd entry

and so on up to

$(r+1)$th entry $= \dfrac{n-r}{r} \times r$th entry

e.g. row 3, 1st entry $= 1$, 2nd entry $= \frac{2}{1} \times 1 = 2$ and 3rd entry $= \frac{1}{2} \times 2 = 1$.

Write algorithms to produce Pascal's triangle using the three different ideas and compare their efficiency.

Sorting algorithms

❓ Working with a partner, take a pack of 52 playing cards each and shuffle them thoroughly. Swap packs and sort the cards into suits in the order clubs, diamonds, hearts, spades, and within each suit from ace up to king as quickly as you can.

Discuss how you went about sorting your cards, and compare your strategies. Try again using the same strategy, or try a different one.

Is a particular strategy always quicker?

Apart from strategy, what other factors will influence the time taken to sort the pack?

Sorting is an everyday activity in which the efficiency of the algorithm used is very important. There are many sorting algorithms designed for use on computers, and there are computer packages available that enable you to compare the merits of the various algorithms.

You will begin by looking at two methods of sorting a list of numbers into ascending order. You will then consider some of the features that you should take into account when comparing them. It should be borne in mind that what might seem an efficient method for a human being may not be so for a computer and vice versa. Both the number of items in the list to be sorted, and how muddled they are, affect the efficiency of the various algorithms.

Selection with interchange sort algorithm

In this algorithm, the smallest number in the list is found and interchanged with the first number. Then the smallest number excluding the first is found and interchanged with the second number. Next the smallest number excluding the first two is found and interchanged with the third number. This process continues until the list is sorted.

To apply this to the list of eight numbers in figure 1.5, look through the whole list to find the smallest number, i.e. 1. Swap this with the 7 in the first position. The result after this first pass through the list is shown alongside.

Now look through the list from the second position downwards to find the smallest number, i.e. 2. Swap this with the 5 in the second position. The resulting list is marked '2nd pass'.

For a list of eight numbers, seven passes will be required before we can guarantee that the list is sorted. The results of the remaining passes are also shown in figure 1.5.

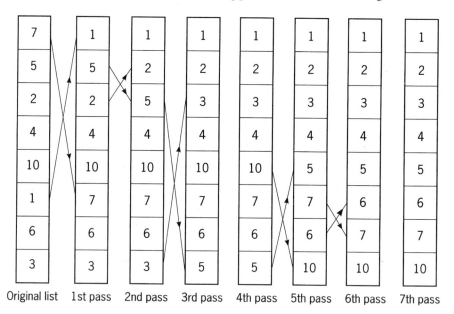

Original list   1st pass   2nd pass   3rd pass   4th pass   5th pass   6th pass   7th pass

Figure 1.5

Note

Although no interchanges take place on the 4th and 7th passes, this is a property of this particular list and the passes must still be carried out to ensure that the list is correctly sorted.

In order to see how this and other sorts can be performed by computer you need to introduce some notation.

Suppose you have the original list shown in figure 1.5.

You can write $L(1) = 7$, $L(2) = 5$ etc. Thus the list is $L(i)$ where i goes from 1 to 8.

The algorithm can now be written as follows:

Look for the smallest number in the list from $L(1)$ to $L(8)$, and swap it with $L(1)$. Look for the smallest number in the list from $L(2)$ to $L(8)$, and swap it with $L(2)$. And so on.

In computer pseudo code this would be:
repeat with $i = 1$ to 7 [look for the smallest number in the list from $L(i)$ to $L(8)$ and swap it with $L(i)$].

When comparing the efficiency of sorting algorithms, two useful measures are the number of comparisons and the number of swaps.

For the selection with interchange sort, you can see that:
the number of comparisons on the 1st pass = 7, 2nd pass = 6, 3rd pass = 5, 4th pass = 4, 5th pass = 3, 6th pass = 2 , and 7th pass = 1.

This gives a total of $7 + 6 + 5 + 4 + 3 + 2 + 1 = 28$ comparisons.
The number of swaps will be at most one per pass, i.e. a maximum of 7.

How many comparisons and swaps would be required for a list of 10 numbers? Try to generalise to obtain formulae for the number of comparisons and swaps when the list is of length *n*.

Bubble sort algorithm

The bubble sort is so named because numbers which are below their correct positions tend to move up to their proper places, like bubbles in a glass of champagne. On the first pass, the first number in the list is compared with the second and whichever is smaller assumes the first position. The second number is then compared with the third and the smaller is placed in the second position, and so on through the list. At the end of the first pass the largest number will have been left behind in the bottom position.

For the second pass the process is repeated but excluding the last number, and on the third pass the last two numbers are excluded. The list is repeatedly processed in this way until no swaps take place in a pass. The list is then sorted.

Starting with the same original list as for the selection with interchange sort, on the first pass we compare 5 and 7, and swap them; then 7 and 2, and swap them; then 7 and 4, and swap them; then 7 and 10, and do not swap them; then 10 and 1, and swap them; then 10 and 6 and swap them, then finally 10 and 3 and swap them. This pass is shown in detail in figure 1.6. Note that the last number is now in its correct position.

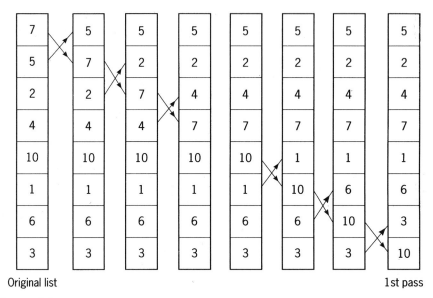

Original list   1st pass

Figure 1.6

The results after making the 2nd and subsequent passes are shown in figure 1.7. You should work through the process to check that you get these results.

<table>
<tr><td>2nd pass</td><td>3rd pass</td><td>4th pass</td><td>5th pass</td><td>6th pass</td></tr>
<tr><td>2</td><td>2</td><td>2</td><td>1</td><td>1</td></tr>
<tr><td>4</td><td>4</td><td>1</td><td>2</td><td>2</td></tr>
<tr><td>5</td><td>1</td><td>4</td><td>3</td><td>3</td></tr>
<tr><td>1</td><td>5</td><td>3</td><td>4</td><td>4</td></tr>
<tr><td>6</td><td>3</td><td>5</td><td>5</td><td>5</td></tr>
<tr><td>3</td><td>6</td><td>6</td><td>6</td><td>6</td></tr>
<tr><td>7</td><td>7</td><td>7</td><td>7</td><td>7</td></tr>
<tr><td>10</td><td>10</td><td>10</td><td>10</td><td>10</td></tr>
</table>

Figure 1.7

The algorithm for the bubble sort for a list of length 8 can be written in computer pseudo code like this:

repeat with $i = 1$ to 7
      [repeat with $j = 1$ to $(8 - i)$
      if $L(j) > L(j + 1)$ swap $L(j)$ and $L(j + l)$]
if no swaps end repeat

The number of comparisons made in a bubble sort for a list of length 8 will be 7 on the first pass, 6 on the second pass, etc. If the maximum number of passes is needed, the total number of comparisons will be $7 + 6 + 5 + 4 + 3 + 2 + 1 = 28$. The number of swaps on the first pass will be anything up to 7; on the second, up to 6, etc. So the maximum possible number of swaps will be: $7 + 6 + 5 + 4 + 3 + 2 + 1 = 28$.

How many comparisons and swaps would be required for a list of size 10? Try to generalise to obtain formulae for the number of comparisons and swaps when the list is of length n.

EXERCISE 1C

1 A common way of writing the date is to use a six-digit number, the first two digits representing the day number, the next two representing the month and the last two representing the last two digits of the year number. Single digit days and months are padded out with leading 0s, thus September 4th 1991 becomes 040991. How would writing the date in year/month/day order (910904 for the example) be advantageous if you wanted to sort events in date order, or to rank a group of people in order of age?

2 Here are two programs in Basic. They use interchange sort and bubble sort, respectively. Investigate which is quicker for different sizes of data sets and varying levels of disorder.

```
FOR i = 1 TO n – 1                          FOR i = 1 TO n – 1
minpos = i                                  swapflag = 0
FOR j = i + 1 TO n                          FOR j = 1 TO n – i
IF L(j) < L(minpos) THEN minpos = j         IF L(j) > L(j + 1) THEN
NEXT j                                      temp = L(j)
temp = L(i)                                 L(j) = L(j + 1)
L(i) = L(minpos)                            L(j + 1) = temp
L(minpos) = temp                            swapflag = 1
NEXT i                                      NEXT j
                                           IF swapflag = 0 THEN i = n – 1
                                           NEXT i
```

The input will be the same for both:

```
READ n
DIM L(n)
FOR i = 1 TO n : READ L(i) : NEXT i
DATA
```

where DATA is the value of n followed by the list of numbers to be sorted. Output can be achieved with:

```
FOR i = 1 TO n : PRINT L(i) : NEXT i
```

As mentioned earlier, computer packages are available that enable you to compare sort algorithms. If you have access to one of these, this would be a good point at which to use it to explore other sort algorithms and carry out a more far-reaching comparison of efficiency. You should be able to consider longer lists and different degrees of initial disorder, thus allowing you to draw more confident conclusions.

The shuttle sort algorithm

1st pass: compare the first two numbers and swap if necessary to place in ascending order.

2nd pass: compare the second and third numbers and swap if necessary, then compare first and second numbers and swap if necessary.

3rd pass: compare the third and fourth numbers and swap if necessary, then compare second and third numbers and swap if necessary and compare first and second numbers and swap if necessary.

And so on. The sequence is shown in figure 1.8.

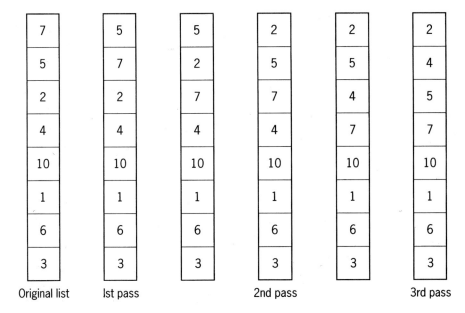

Original list   1st pass   2nd pass   3rd pass

Figure 1.8

Insertion sort algorithm

This is similar to the way you might arrange a hand of cards. Numbers are taken one at a time, in sequence, from the original list and inserted in their correct positions in a new list as shown in figure 1.9.

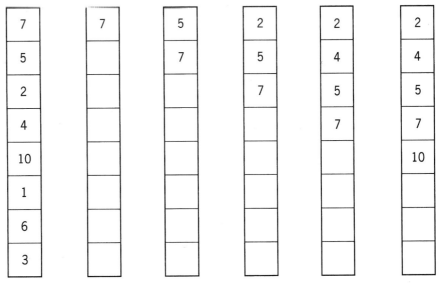

Original list

Figure 1.9

The quick sort algorithm

First split the list into two sub-lists, one containing those numbers less than or equal to the first number in the list, the other containing those greater than it. Do not reorder the sub-lists. Place the first number between the two sub-lists. Repeat the process on sub-lists containing two or more numbers until there are no such sub-lists. The list is then sorted. The process is shown in figure 1.10.

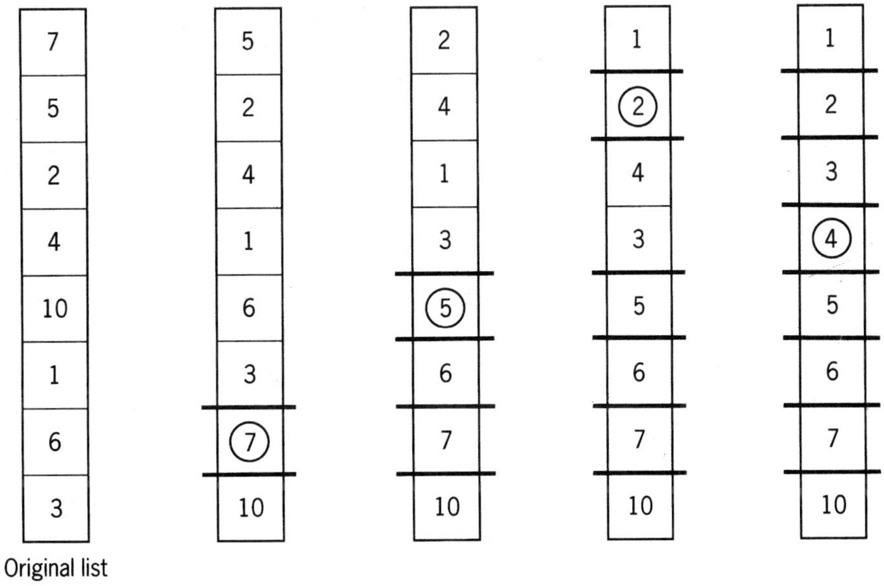

Original list

Figure 1.10

Look carefully at the details of the shuttle sort, insertion sort and quick sort algorithms. Carry out the following investigation on them.

(i) Apply them to some lists of numbers (or words) to enable you to see clearly how they work.

(ii) Write the algorithms in either pseudo code or some suitable programming language, and test them if you have access to a computer.

(iii) Assess their efficiency by counting the number of comparisons and swaps that will be involved. Try to obtain generalisations where possible, and compare your results with those for the selection with interchange sort and the bubble sort algorithms.

Searching

ACTIVITIES

1 Working in pairs, tell your partner to think of a whole number between 1 and 99 (inclusive). Then by asking suitable questions to which the answer must be either yes or no, try to discover the number. The aim is to find the number by asking as few questions as possible. Swap over and let your partner try to find a number that you are thinking of. Repeat the exercise several times, trying different strategies. Discuss your strategies. What is the minimum, maximum and average number of questions required?

2 Treasure Island. (Again you will need a partner.)

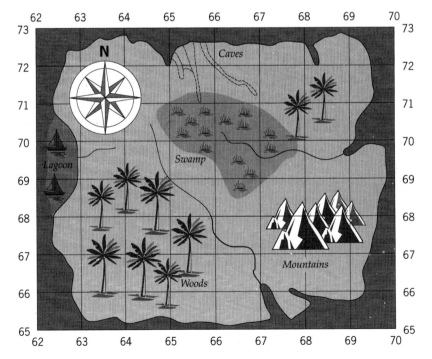

Figure 1.11

Somewhere on the island shown on the map (figure 1.11) is buried a chest of pirate treasure. Ask your partner to decide on the six-figure grid reference of where it is hidden. You are allowed nine attempts at guessing the grid reference of the location of the treasure. After each attempt your partner will tell you if you need to go north or south and if you need to go east or west. Take it in turns to 'bury' the treasure and when you have played the game a few times discuss your strategies.

3 Use a dictionary or telephone directory to investigate how many items of data you look at when searching for a given word or someone's phone number. Compare various strategies and try to describe them as algorithms.

Searching, like sorting, is an everyday activity that has attracted a lot of attention from the writers of algorithms. In today's information-rich society, finding the item of data that you require can be like looking for a needle in a haystack. You will now consider three methods.

Linear search algorithm

This is the simplest of the search algorithms, in which you check each item of data in turn to see if it satisfies your criterion. There are no restrictions on the data; it will work even if the data are not ordered. It is, however, most inefficient. If the item for which you are looking is not there, you will still have checked every item of data. Imagine trying to find the name of a person, given their telephone number, by searching the telephone directory!

It is usually worth ordering the data in a way that suits your requirements. Often the same set of data will be ordered in different ways to facilitate different types of search request. For example, a library catalogue is ordered both according to author and according to book title. If the data are ordered there are two algorithms that we can consider using.

Indexed sequential search algorithm

The data are first ordered and then subdivided. An extra list or index is then created containing the first or last item in each subdivision. Such a method is used in a dictionary where the index is positioned at the top right-hand corner of the page. To find a given word, you first leaf through the pages, looking at the index, to locate the page that the word is on. Then you carry out a linear search on the selected page.

For a set of data held on a computer system you would need to create a sub-list. For example if you had a list of the telephone area codes for the UK ordered by town name, the sub-list could contain the position of the first town whose name began with A, B, C, ... etc. Thus to find York, the sub-list would first be searched to find Y. This would give the position from which to start the linear search of the main data.

Binary search algorithm

The data are first sorted into ascending order. The following steps are then carried out:

Step 1   Look at the middle item.
            If this is the required item the search is finished.
            If not, the item is in either the top or bottom half: decide which half by comparison with the middle item.
Step 2   Apply Step 1 to the chosen half.

At each stage the number of items to be searched is halved, hence the name of the algorithm.

---

❷ The algorithm proceeds most smoothly when the number of data items is 3, 7, 15, 31 or 63, etc. (i.e. of the form $2^n - 1$). Can you see why? What will be the maximum number of comparisons for each of these numbers of data items?

It can be worth adding dummy items to give a total of the form $2^n - 1$, but if this leads to adding a lot of extra items, it becomes preferable to adapt the algorithm to overcome the difficulty with the middle item.

INVESTIGATION

Investigate

(i)   the maximum number of comparisons per request

(ii) the mean number of comparisons per request

when searching a list of n items for a specific item of data using a linear search algorithm.

Repeat the investigation, this time for a binary search algorithm.

You may assume in each case that the list is sorted into ascending order. If you are unable to obtain general expressions, you should give results for a selection of values of n.

1 Using the notation L(1)... L(n) for the list of n items, write algorithms in either pseudo code or a computer programming language for

(i) the linear search

(ii) the binary search.

If you have used a programming language and have access to a computer, test your results.

2 If you have access to a large amount of data in computer form (e.g. a list of pupils in your school with their dates of birth), write an indexed sequential search algorithm to find individual pupils' dates of birth.

3 Suppose you wanted to write a computer program to play the number guessing game in the introductory exercise. What algorithm would you use to enable the computer to guess your number? Try to write a suitable program.

4 The following algorithm finds the highest common factor (H.C.F.) of two positive integers (e.g. the H.C.F. of 24 and 36 is 12).

1   Let A be the first integer and B be the second integer.
2   Divide B by A and round down to the nearest integer. Let Q be this result.
3   Let $R = B - (Q \times A)$.
4   If $R = 0$ go to step 8.
5   Let the new value of B be A.
6   Let the new value of A be R.
7   Go to step 2.
8   Record the H.C.F. as the value of A.
9   Stop.

(i) Work through the algorithm with $A = 2520$ and $B = 5940$.

Iteration	A	B	Q	R
1	2520	5940		

(ii) What happens if the order of input is reversed, i.e. if $A = 5940$ and $B = 2520$?

(iii) It is claimed that the number of iterations of this algorithm is approximately

$$\frac{\log[M/1.17]}{\log[(1 + \sqrt{5})/2]} \text{ , where } M \text{ is the larger of } A \text{ and } B.$$

Investigate the number of iterations of the loop in the algorithm when $A = 233$ and $B = 377$.

Compare with the prediction from the formula.

[Oxford]

5 The following flowchart defines an algorithm which operates on two inputs, x and y.

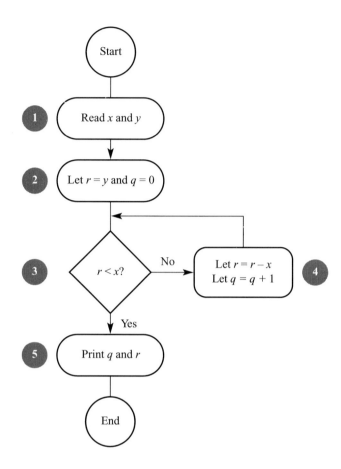

(i) Run the algorithm with inputs of $x = 3$ and $y = 41$, counting how many times the instructions in box number 4 are repeated.
Record the printed value of q, the printed value of r and the number of repetitions of box number 4.

(ii) Say what the algorithm achieves.

The following flowchart defines an algorithm with three inputs, x, y_1 and y_2.

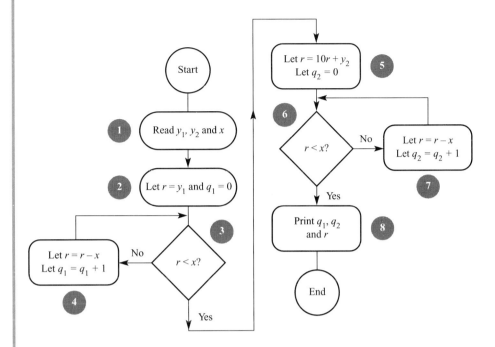

(iii) Work through the algorithm with $x = 3$, $y_1 = 4$ and $y_2 = 1$. Record your work by copying and completing the table below. Keep a count in the spaces provided of how many times the instructions in boxes numbers 4 and 7 are repeated.

r																								
q_1																								
q_2																								
No. of repetitions of box number 4																								
No. of repetitions of box number 7																								

Give the printed values of q_1, q_2 and r.

(iv) The second algorithm achieves the same result as the first. Say what you think are the advantages and disadvantages of each.

[MEI]

6 (i) The instructions labelled 1 to 7 below describe the steps of a bubble sort algorithm. Apply the bubble sort algorithm to the list of numbers shown below. Record in a table the state of the list after each pass, and record the number of comparisons and the number of swaps that you make in each pass. (The result of the first pass has already been recorded.)

Instructions

1   Let i equal 1.
2   Repeat lines 3 to 7, stopping when i becomes 6.
3   Let j equal 1.
4   Repeat lines 5 and 6, stopping when j becomes $7 - i$.
5   If the jth number in the list is bigger than the $(j + 1)$th, then swap them.
6   Let the new value of j be $j + 1$.
7   Let the new value of i be $i + 1$.

List: 7   9   5   1   11   3

Table *i*th pass	$i=1$	$i=2$	$i=3$	$i=4$	$i=5$
	7				
	5				
	1				
	9				
	3				
	11				
Comparisons	5				
Swaps	3				

(ii) Suppose now that the list is split into two halves, {7, 9, 5} and {1, 11, 3}, and that the algorithm is applied to each half separately, giving {5, 7, 9} and {1, 3, 11}.

How many comparisons and swaps does this entail altogether?

(iii) A modified version of the bubble sort algorithm, in which lines 2 and 3 have been changed, is now used to sort the revised list {5, 7, 9, 1, 3, 11}.

Revised instructions

1   Let i equal 1.
2   Repeat lines 3 to 7, stopping when i becomes 4.
3   Let j equal $4 - i$.
4   Repeat lines 5 and 6, stopping when j becomes $7 - i$.

5   If the jth number in the list is bigger than the $(j+1)$th, then swap them.

6   Let the new value of j be $j + 1$.

7   Let the new value of i be $i + 1$.

Apply the revised algorithm to the list $\{5, 7, 9, 1, 3, 11\}$ recording your results in a table.

Does splitting the list, applying the bubble sort algorithm to each half separately, and then applying the revised algorithm to the revised list, appear to offer any advantages?

Justify your answer.

(iv) Explain why the changes to statements 2 and 3 of the original algorithm were made to produce an algorithm to sort the revised list.

[**MEI**]

7 (i) The following eight lengths of pipe are to be cut from three pieces, each of length 2 m:

1.1 m, 1.2 m, 0.3 m, 0.4 m, 0.2 m, 0.4 m, 0.3 m, 0.7 m.

(a) Use the first-fit decreasing algorithm to find a plan for cutting the lengths of pipe. Show your plan by marking cuts on a copy of the diagram below.

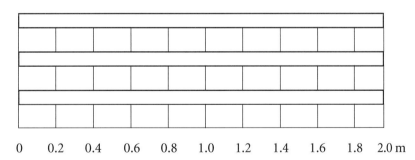

0   0.2   0.4   0.6   0.8   1.0   1.2   1.4   1.6   1.8   2.0 m

(b) Suppose that it is preferable to be left with a small number of longer pieces of pipe, rather than a larger number of shorter lengths. Use any method to produce an improved cutting plan as below.

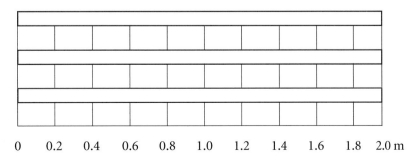

0   0.2   0.4   0.6   0.8   1.0   1.2   1.4   1.6   1.8   2.0 m

(ii) Here is an algorithm for placing a number of rectangular tiles into a large rectangular grid without overlap:

1   Take the first tile.
2   Starting from the top left, choose the first vacant square of the large rectangle, searching for it along the top row, then the second row, etc.
3   If possible, place the tile 'lengthways on' (ie. ▭), so that its top-left corner corresponds to the top-left corner of the chosen grid square. If it can be placed then go to instruction 6.
4   If the tile cannot be placed 'lengthways on' then try it 'end-on' (ie. ▯) so that its top-left corner corresponds to the top-left corner of the chosen grid square. If it can be placed then go to instruction 6.
5   Choose the next vacant square to the right if there is one and go to instruction 3.

    If there are no more vacant squares in the current row then choose the leftmost vacant square in the next row down and go to instruction 3.

    If there are no more rows then go to instruction 7.
6   Take the next unplaced tile from the list and go to instruction 5.
    If there are no tiles left to place go to instruction 8.
7   Report that the task has not been completed and stop.
8   Report that the task has been completed and stop.

(a) Apply the algorithm to attempt to fit the following six tiles into an 8×4 rectangle:

Tile Number	Length	Breadth
1	4	1
2	3	2
3	5	2
4	4	1
5	2	1
6	3	2

Draw the positions of the tiles on a copy of the diagram as they are placed:

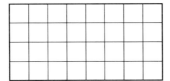

(b) You should have found that the last three tiles are not placed by the algorithm.

Change instruction 6 as follows:

6 Take the next unplaced tile from the list and go to instruction 2. If there are no tiles left to place go to instruction 8.

Explain why the revised algorithm, when applied to the list of tiles in (ii)(a), leaves only one tile unplaced.

(c) Describe a further improvement of the modified algorithm so that all six of these tiles are placed.

(d) Show the result of applying your algorithm from part (c) on a copy of the diagram.

[MEI]

8 This question is concerned with shuttle sort and insertion sort.

Shuttle sort

First pass   Compare the first two numbers and swap if necessary so that the smaller is written first.
Write down the remaining numbers with their order unchanged.

Second pass In the revised list compare the second and third numbers and swap if necessary.
If a swap was necessary then compare the revised first and second numbers and swap if necessary.
Write down the remaining numbers with their order unchanged.

Third pass   In the revised list compare the third and fourth numbers and swap if necessary.
If a swap was necessary then compare the revised second and third numbers and swap if necessary.
If both swaps were necessary then compare the revised first and second numbers and swap if necessary.
Write down the remaining numbers with their order unchanged.

And so on.

Insertion sort

Stage 1   Write down the first number on the list.

Stage 2   Write down the next number before the first if it is lower, or after if it is greater.

Stage 3   Take the next number on the list of numbers to be sorted and compare it to the first number on the written list. Write it down before it if it is less. If not then compare it to the next number on the written list. Write it before it if it is less, or after it if it is greater.

Stage 4   Take the next number on the list of numbers to be sorted and compare it to the first number on the written list. Write it down before it if it is less. If not then compare it to the next number on the written list. Write it before it if it is less. If not then compare it to the next number on the written list. Write it before it if it is less, or after it if it is greater.

And so on.

(i) **(a)** Use shuttle sort to sort the following list. Write down the list as it appears at the end of each pass, and count the number of comparisons that you have to make within each pass.

List = {13, 56, 2, 40, 10, 50, 35}

(b) Use insertion sort to sort the same list. Show the written list at each of the stages and count the number of comparisons that you have to make at each stage.

List = {13, 56, 2, 40, 10, 50, 35}

(ii) The list {3, 1, 2} consists of the first three natural numbers in a particular order. This is known as a *permutation* of the numbers {1, 2, 3}.

There are six permutations of the numbers {1, 2, 3}. Say how many comparisons each of shuttle sort and insertion sort would need to make to sort them.

(iii) (a) Find a permutation of the set {1, 2, 3, 4, 5, 6, 7} which leads to the largest number of comparisons when sorted by shuttle sort. Give the number of comparisons.

(b) Find a permutation which leads to the largest number of comparisons when sorted by insertion sort. Give the number of comparisons.

(c) Without doing a sort, suggest the largest number of comparisons needed to sort a list of length 10 by either shuttle sort or insertion sort.

[MEI]

9 (i) In a two person game one player thinks of a word and invites the other to guess what that word is. In response to an incorrect guess the first player indicates whether the mystery word is before or after the guess in alphabetical (dictionary) order.

    (a) Given that a dictionary is available, describe a strategy for the second player to follow in trying to find the word in as few guesses as possible.

    (b) Given that the dictionary has approximately 100 000 words, find the maximum number of guesses which might be required to find the mystery word using your method from part (a). Explain your reasoning in arriving at this number.

(ii) In a game of 'battleships' a player chooses a square on a grid to represent a battleship. A second player tries to find that square by making a sequence of guesses. In response to an incorrect guess, the second player is told **either** whether the battleship is to the east or west of the guess (i.e. to the left or the right), **or** whether it is to the north or south (above or below).

The game is being played on a 15×20 grid and a first guess has been made at square K8, as indicated on the diagram below.

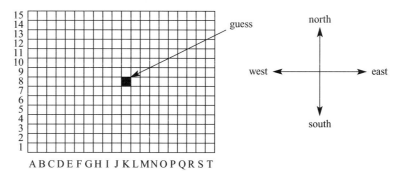

The information provided is that the battleship is to the south of the guess.

    (a) In a binary search the set of possibilities is, as nearly as possible, halved at each iteration. Given the guess and the information, what would be the next guess?

    (b) If the information following that next guess is that the battleship is to the north of the guess what would be the next guess in a binary search?

    (c) What difficulty will be encountered when the information is that the battleship is to the west?

    (d) Starting a game afresh, what is the minimum number of guesses that will be needed to be sure of locating the battleship?

(iii) 'Spaceships' is similar to 'battleships', but is played in a three-dimensional grid.
Playing a binary search strategy on a $15 \times 15 \times 15$ grid, how many guesses will be needed to guarantee locating the spaceship?

[MEI]

10 (i) An algorithm for sorting a list of different numbers into ascending order is described below.

1 Write down the first number in the list and circle it to show that it is not to be examined again.

2 Compare the next number in the list with the circled number and write it before the circled number if it is less, or after if it is greater.

3 Compare the next number in the list with the circled number and write it before or after, as in step 2. Write it after any other number that has already been dealt with in this part of the algorithm. (Note that this means that only one comparison is made – with the circled number.)

4 Continue until the last element in the list has been dealt with. Note that the process will have created two sub-lists, one before the circled number and one after, either of which could be empty.

5 Continue to apply the process to sub-lists of length two or more.

(a) Use this algorithm to sort the list {13, 56, 2, 40, 10, 50, 35}. Show all of your steps and count the number of comparisons that you have to make.

(b) Find the number of comparisons required to sort the list {7, 6, 5, 4, 3, 2, 1} into ascending order using this algorithm.

(ii) In a game of hide-and-seek Mary has to start from A and inspect hiding places at B, C, D, E, F and G. The routes which are available are shown in the diagram. The edges are all of length 1.

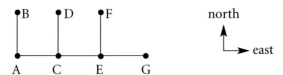

Let hiding places which are at a distance 1 from A be called **level 1** places, hiding places which are at distance 2 from A be called **level 2** places, etc.

Mary uses the following algorithm to inspect all of the hiding places.

● Inspect all hiding places at level n before looking at places at level $n+1$.
● When there is a choice of search direction go east before north.

Thus she starts her inspection as follows: A→C→A→B→A→C→E→...

(a) Complete Mary's route and give its total length.

An alternative approach for Mary is given by the following algorithm.

● Where possible move to a hiding place which has not yet been inspected.
● When there is a choice of direction go east before north.

(b) Give Mary's route and its length under this revised algorithm.

(c) Write an algorithm which will produce a route of minimum length for Mary.
Give the route and its length.

[MEI]

11 Each office in a block of offices has its own filing system. The systems consist of wallets in which documents are placed.

However, wallets can also be placed inside other wallets.

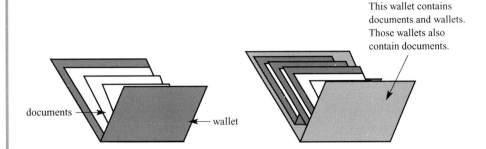

This wallet contains documents and wallets. Those wallets also contain documents.

documents — wallet

In one office the arrangement of wallets within wallets can be represented graphically as follows.

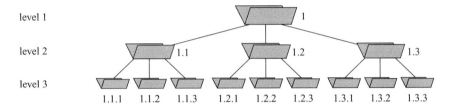

level 1   1

level 2   1.1   1.2   1.3

level 3   1.1.1   1.1.2   1.1.3   1.2.1   1.2.2   1.2.3   1.3.1   1.3.2   1.3.3

The system consists of a wallet at *level 1* (labelled wallet 1) which contains three *level 2* wallets. These are labelled as shown. Each of these level 2 wallets contain, in turn, three *level 3* wallets, labelled so that wallet 1.3.2, for example, is the second wallet contained in wallet 1.3.

Each wallet also contains documents.

(i) (a) In another office the filing system is similar, with three level 2 wallets inside wallet 1, and with three level 3 wallets inside each level 2 wallet. But in addition there are three level 4 wallets inside each level 3 wallet.

State how many wallets there are in total in this system.

(b) State how many wallets there would be in total if the system were to be extended in the same way to five levels.

(ii) In the first office a document has to be retrieved from the three-level system. The following algorithm (called *depth first search*) is to be used.

1   Call wallet 1 the *current wallet*.
2   Examine the current wallet to see if the required document is there. If it is not then go to step 3.
3   Let the first unexamined wallet (in numerical order) in the current wallet become the new current wallet, and go to step 2. If there is no unexamined wallet then go to step 4.
4   Let the wallet which contains the current wallet become the new current wallet, and go to step 3.

If the required document is in wallet 1.3.1, list the wallets that are examined, in the order in which they are examined, in using depth first search to find the document.

(iii) An alternative algorithm is called *breadth first search*. In this algorithm all of the wallets at level n are examined before any wallet at level $n + 1$ is examined. At any level wallets are to be examined in numerical order.

If the required document is in folder 1.3.1, list the wallets that are examined, in the order in which they are examined, in using breadth first search to find the document.

(iv) For each of the wallets listed below place a tick in one row of a copy of the table to show whether it is examined earlier under breadth first search than under depth first search, whether later, or whether there is no difference.

	1.1	1.2	1.3	1.1.1	1.1.2	1.1.3	1.2.1	1.2.2	1.2.3	1.3.1	1.3.2	1.3.3
Examined earlier with breadth first search												
Examined later with breadth first search												
Wallets for which there is no difference												

(v) An alternative way of comparing the two search algorithms is to count the number of documents which have to be examined before the required document is found. Explain why, using this approach, neither of the methods is inherently better than the other.

[MEI]

12 The ferry *Bougainvillea* runs a service between two islands. The vehicle deck has three lanes with lengths as shown in the diagram.

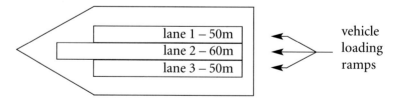

(i) Cars and commercial vehicles wait for the ferry in a single queue along the quayside. The queue for the 8 am sailing consists of vehicles which will take up the lengths of lane indicated in the table. (The lengths are given in the order in which the vehicles are queueing.)

Vehicle	1	2	3	4	5	6	7	8	9	10	11	12	13	14
Length	3	4	4	3.5	8	9	3.5	10	3	3	4	4	3	8

Vehicle	15	16	17	18	19	20	21	22	23	24	25	26	27	
Length	11	3	10	10	10.5	3.5	8	8	4	4.5	4.5	8	4	

(a) The following algorithm is used to load vehicles.

> *Direct the vehicles into lane 1 until the next vehicle to be loaded cannot be placed in lane 1. Then direct the vehicles into lane 2 until the next vehicle to be loaded cannot be placed in lane 2. Then direct the vehicles into lane 3 until the next vehicle to be loaded cannot be placed in lane 3. Then cease loading.*

Using this algorithm give the total lengths used in each lane and state how many of the 27 vehicles are loaded.

(b) Give a simple improvement to this lane-by-lane approach. Say what difference it will make to loading this queue of vehicles.

(c) An alternative loading algorithm is:

> *Load the next vehicle into the lane with the maximum length remaining, choosing the lowest numbered lane in the event of a tie. Cease loading when the next vehicle cannot be loaded.*

Show the result of this algorithm by completing a copy of the diagram.

vehicle number

4										lane 1
46.5										

length remaining

vehicle number

1	2	3								lane 2
57	53	49								

length remaining

vehicle number

										lane 3

length remaining

(d) Suggest what might be done to improve this approach.
(You are not required to give details of how to change the algorithm.)

(ii) It is proposed that the single queue be replaced by two parallel queues. Vehicles greater than 5 m in length will use the first queue, the second queue being reserved for vehicles of length 5 m or less.

(a) The algorithm in part (i)(c) is modified as follows:

First load the next vehicle from queue 1 (long vehicles) into the lane with the maximum length remaining, choosing the lowest numbered lane in the event of a tie.
When all vehicles from queue 1 are loaded repeat for vehicles from queue 2.

Show the results of this algorithm by completing a copy of the diagram below.

vehicle number
length remaining

lane 1

vehicle number
length remaining

lane 2

vehicle number
length remaining

lane 3

(b) Suggest a practical disadvantage in using this two-queue system.

[MEI]

13 Consider the following algorithm:

```
1    INPUT n
2    LET J = 1
3    REPEAT
4        LET X(J) = 0
5        LET J = J + 1
6    UNTIL J = n + 1
7    LET J = 1
8    REPEAT
9        LET K = J
10       REPEAT
11               LET X(K) = 1 – X(K)
12               LET K = K + J
13           UNTIL K > n
14       LET J = J + 1
15   UNTIL J = n + 1
```

(i) The input value, n, must be a positive integer.

Run the algorithm for various values of n, including $n = 17$, and report on what it achieves.

(ii) Explain why it works.

[Oxford]

1 An algorithm is a well-defined, finite sequence of instructions used to solve a problem.

2 Algorithms can be communicated in various ways, including

- written english
- pseudo code
- computer programming languages
- flowcharts
- structure diagrams.

3 The efficiency of an algorithm is measured in terms of its complexity and its data storage requirements.

4 A heuristic algorithm *is* an algorithm but it does not guarantee an optimal solution to the problem.

5 There are various types of sorting algorithm. To compare their efficiencies you need to consider

- the number of comparisons
- the number of interchanges.

6 Systematic search methods include linear, indexed sequential and binary. The efficiency of a search method depends on the percentage of the data, on average, that needs to be examined before you find the item you want.

Graphs

A picture is worth a thousand words.

Frederick R. Barnard

The following is an ancient (medieval) puzzle.

A showman needs to carry across a river his wolf, his goat and a sack of cabbages. Left unsupervised the goat would eat the cabbages. Left unsupervised the wolf would eat the goat. There is a boat available, but it is small and can accommodate only the showman, who must row, plus one other of the three.

Suppose that you use the symbol SG|WC to represent the situation in which the showman and the goat are on the first bank (with the boat), and the wolf and the cabbages are on the far bank, having been ferried across. There are 16 such symbols, since each letter has to appear either to the left or the right of the '|', but 6 of the 16 are 'illegal' (in that they would lead to some eating).

 1 What do the lines linking the symbols mean?
2 What is figure 2.1 saying?

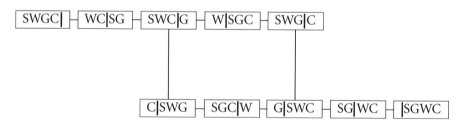

Figure 2.1

You should be able to see that figure 2.1 shows that there are two ways in which the showman can get the entire party safely across by making seven crossings (and many more ways involving more than seven crossings!).

Figure 2.1 is a *graph*. The symbols are *vertices* (or *nodes*). The lines are *edges* (or *arcs*).

Background

If you were asked to 'draw a graph', you might well produce something like figure 2.2.

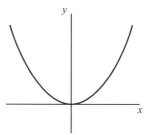

Figure 2.2

This is a very powerful way of visualising a mathematical relationship, and a useful tool for solving problems. However, it is only one type of graph. Others appear in different circumstances to solve different types of problems.

For example figure 2.3 is a graph.

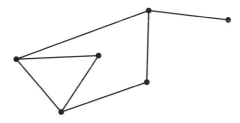

Figure 2.3

Graphs like this one can be used to describe and solve many different types of problem that involve *discrete sets*. They do not have axes; they are simply lines (called *edges* or *arcs*) connecting points (called *vertices* or *nodes*). At university level there is a whole branch of mathematics, called *graph theory*, devoted to the study of this type of graph. In this course you will use such graphs as a powerful way of describing the relationships between members of *discrete sets*.

Figure 2.1 is such a graph. The nodes are 'legal' (no eating) configurations of the showman, the goat, the wolf and the cabbages. The nodes are linked by arcs when it is possible, with one ferry trip, to get from the situation described by the node at one end of the arc to the situation described by the node at the other end.

Figure 2.2 is the type of graph that comes to mind when you are working with real numbers. It describes the relationship between the real numbers on the x axis and those on the y axis. (It looks as if the relationship might be something like $y = x^2$.) Real numbers are *continuous*. (Given *any* two real numbers, you can always find a third real number between them. Just add them together and divide the result by 2.) Mathematics dealing with real numbers is known as *continuous mathematics*.

However, in many situations you have to deal with *discrete sets*, and so continuous mathematics is not appropriate. A discrete set is one which can be put in one-to-one correspondence either with the positive integers, or with a subset of the positive integers. In everyday language, it can be *counted*.

The distinction between discrete and continuous is fundamental. Discrete sets are counted. Continuous sets are measured, and are associated with words such as weight, length, amount, etc.

Next time you hear the phrase '... amount of people ...' you should realise that the user of the phrase does not have the distinction clear in his or her mind!

EXAMPLE 2.1

$X = \{\text{London, Oxford, Birmingham, Cambridge, Leicester}\}$

Let $X \times X$ be the set of all possible pairs from the set X. Since the set X has 5 members, $X \times X$ must have 25 members, such as

(London, Oxford)
(Oxford, London)
(London, London), etc.

Consider the graph consisting of all of those elements of $X \times X$ which represent direct motorway routes. ('Direct' means not via any of the other listed cities).

This would consist of

$\left\{ \begin{array}{lll} \text{(London, Oxford),} & \text{(London, Birmingham),} & \text{(London, Leicester),} \\ \text{(London, Cambridge),} & \text{(Oxford, London),} & \text{(Oxford, Birmingham),} \\ \text{(Birmingham, Oxford),} & \text{(Birmingham, London),} & \text{(Birmingham, Leicester),} \\ \text{(Leicester, London),} & \text{(Leicester, Birmingham),} & \text{(Cambridge, London)} \end{array} \right\}$

You could represent this set by the following picture (figure 2.4) which is usually called the *graph*.

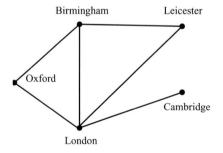

Figure 2.4

Figure 2.4 has been drawn so that the cities all are roughly in the correct geographical position, but it need not have been. Figure 2.5 is exactly the same graph – it conveys exactly the same information.

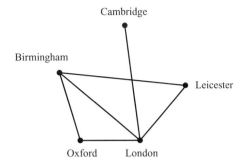

Figure 2.5

The lines (called edges or arcs) do not represent motorways – they represent the relationship 'is connected directly by motorway'. Indeed a *map* of the motorways would look more like figure 2.6.

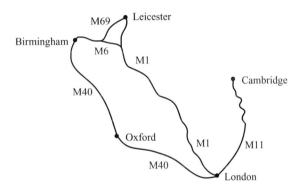

Figure 2.6

EXAMPLE 2.2

$X = \{2, 3, 4, 5, 6\}$

The graph in figure 2.7 illustrates (is) the relationship 'share a common factor other than 1'.

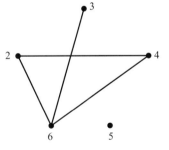

Figure 2.7

Graphs of all types are very useful in organising and solving problems.

 Can you join these points to make this without taking your pen off the paper or repeating a line?

B
A ● ● C
E ● ● D

B
A ● ● C
E ● ● D

Question 7 in Exercise 2A deals with this type of problem.

Definitions

The following definitions refer to the pictorial representation of a graph on $X \times X$, where X is a finite discrete set.

- A graph is a set of *vertices* (or *nodes*), one for each member of X, together with a set of *edges* (or *arcs*).

- An *edge* has a vertex at each end. An edge with the same vertex at each end is called a *loop* (see figure 2.8).

Figure 2.8

- The *degree* (or *order*) of a vertex is the number of edges incident on it.

a 4 node

Figure 2.9

- A *simple graph* is one in which there are no loops, and in which there is no more than one edge connecting any pair of vertices.

- A *walk* is a sequence of edges in which the end of one edge (except the last) is the beginning of the next.

- A *trail* is a walk in which no edge is repeated.

- A *path* is a trail in which no vertex is repeated (see figure 2.10).

Figure 2.10

- A *cycle* is a closed path, i.e. the end of the last edge is the start of the first (see figure 2.11).

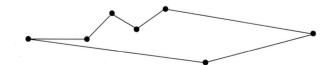

Figure 2.11

- A *Hamiltonian cycle* is a cycle which visits every vertex. (Note that since a Hamiltonian cycle is a cycle, and since a cycle is a path, each vertex is visited once and only once.)

- A graph is *connected* if there exists a path between every pair of vertices.

- A *tree* is a simple connected graph with no cycles (see figure 2.12).

Figure 2.12

- A *digraph* (directed graph) is a graph in which at least one edge has a direction associated with it (see figure 2.13). (In our motorway example such edges would be needed if any one-way motorways existed!)

Figure 2.13

- A *complete graph* is a simple graph in which every pair of vertices is connected by an edge (see figure 2.14).

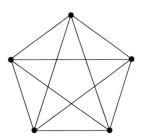

Figure 2.14

- An *incidence matrix* is a way of representing a graph by a matrix. For example, in figure 2.15, the graph is represented by the matrix next to it.

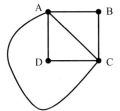

$$\begin{array}{c} \\ A \\ B \\ C \\ D \end{array} \begin{array}{cccc} A & B & C & D \\ \begin{bmatrix} 0 & 1 & 2 & 1 \\ 1 & 0 & 1 & 0 \\ 2 & 1 & 0 & 1 \\ 1 & 0 & 1 & 0 \end{bmatrix} \end{array}$$

Figure 2.15

- Two graphs are *isomorphic* if one can be stretched, twisted or otherwise distorted into the other. In figure 2.16, graphs 1 and 2 are isomorphic to one another, but graph 3 is not isomorphic to them.

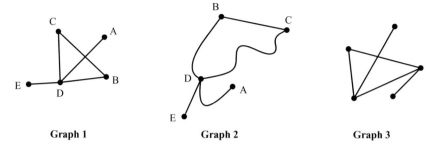

| Graph 1 | Graph 2 | Graph 3 |

Figure 2.16

Note that when graphs are isomorphic it is possible to label their vertices so that the linkage between corresponding vertices is the same. Graphs 1 and 2 have had their vertices labelled to show they are isomorphic. It is not possible to achieve an equivalent labelling on graph 3.

● A *planar graph* is one which can be drawn without any edges crossing.

● A *bipartite graph* is one in which the vertices fall into two sets and in which each edge has a vertex from one set at one end and from the other set at its other end.

This example demonstrates how a graph can be used to help to solve a problem.

A 90 kg man, a 50 kg woman and a 40 kg child have a pair of buckets connected with rope over a pulley to lower themselves to the ground. A 30 kg sack of potatoes is also available. No more than two of the four can occupy any one bucket. If a man, woman or child is in any bucket then there must be no more than a 10 kg difference in the load in each bucket, or a nasty accident will occur.

Figure 2.17

SOLUTION

You can illustrate the possible situations by drawing a 16 vertex digraph, each vertex representing who/what is still at the top, and hence who/what has reached the ground (figure 2.18). You can then start to connect vertices if it is possible to move safely from the situation represented by one vertex to another. If you can

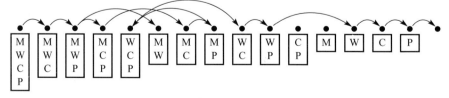

Figure 2.18

find a path from the first vertex to the last then you will have solved the problem. Not all of the edges of the graph are shown here, but a path now exists (figure 2.19).

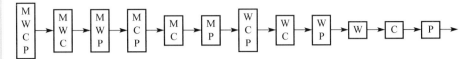

Figure 2.19

❓ Try writing a set of instructions to implement the solution represented by this graph.

*Questions marked * may be suitable starting points for coursework.*

1 The table below shows the numbers of vertices of orders 1, 2, 3 and 4 in four different graphs. Draw an example of each of these graphs.

Order of vertex	1	2	3	4
Graph 1	4	0	0	1
Graph 2	0	0	4	1
Graph 3	0	1	0	1
Graph 4	2	0	0	1

2 **(i)** List all of the cycles in the graph below which can start and finish at A. Note that, for example, ABCA and ACBA represent the same cycle.

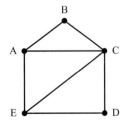

(ii) Why is ABCEDCA not a cycle?

(iii) What could ABCEDCA best be described as?

3 This is the incidence matrix for a graph. Draw the graph it represents.

$$\begin{array}{c c c c} & A & B & C \\ A & \begin{bmatrix} 0 & 1 & 2 \\ B & 1 & 2 & 1 \\ C & 2 & 1 & 0 \end{bmatrix} \end{array}$$

4 Prove that, in a graph, the number of vertices of odd degree is even. (Note that a loop contributes 2 to the degree of its vertex.)

5 If G1 is simply connected and has four vertices, what is the least number of edges it could have, and what is the greatest number of edges it could have? If G2 is simply connected and has *n* vertices, what is the least number of edges it could have, and what is the greatest number of edges it could have?

[AEB]

6* (You may recognise this problem. Look back to investigation 4 on page 14.) The weights and values of five items are as given in the table:

Item	A	B	C	D	E
Weight (kg)	3	8	6	4	2
Value	2	12	9	3	5

Items are to be selected whose total value is as large as possible, subject to the constraint that the total weight must not exceed 9 kg. No item may be selected twice.

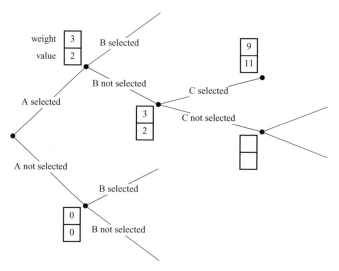

The problem is to be solved by constructing a tree whose branches represent selecting or not selecting an item:

(i) Copy and complete the tree recording the cumulative weight and value of items selected so far at each vertex. Do not continue along a branch after the cumulative weight reaches 9 kg or more. Hence give the best solution to the problem.

(ii) Imposing the condition that branches are not continued after the cumulative weight reaches 9 kg is known as a *bound*. How many more vertices would you have needed on your diagram had there been no bound on the total weight?

(iii) The following two graphs are not trees. For each of these graphs say why it is not a tree and explain why it could not be used as a model, or as part of a model, for a similar selection problem.

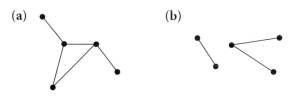

(a)   (b)

[AEB]

7* In the 18th century the inhabitants of Königsberg, now Kaliningrad, enjoyed promenading across the town's seven bridges. It was known not to be possible to cross each bridge once and only once.

Leonhard Euler (1707–83) translated the problem into graph theory, and generalised the result.

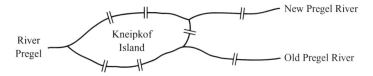

(i) Explain how the following graph models the Königsberg problem.

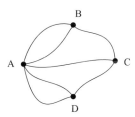

(ii) A graph in which it is possible to trace each edge once and only once without lifting pencil from paper, returning to the start, is called *Eulerian*, or traversable. Euler's result was that a connected graph is traversable if and only if it has no odd vertices.

Explain why a graph with odd vertices cannot be traversable.

(iii) A connected graph with just two odd vertices is called *semi-Eulerian*. Show that it is possible to trace each edge of such a graph once and only once without lifting pencil from paper.

8 **(i)** A simple connected graph has seven vertices, all having the same degree d. Give the possible values of d, and for each value of d give the number of edges of the graph.

(ii) Another simple connected graph has eight vertices, all having the same degree d. Draw such a graph with $d = 3$, and give the other possible values of d.

[AEB]

9 The caller at a barn dance asks four males and four females to arrange themselves so that the females are each holding hands with two other people, whilst the males are each holding hands with just one other person.

Draw graphs using ● to represent a female and × to represent a male, to show all possible different arrangements in which this can be achieved.

For example:

You are not required to consider the number of ways in which arrangements may be formed. In other words you should regard the males as being indistinguishable from each other, and likewise for the females.

[AEB]

10 A puzzle has four cubes with four different colours on their faces. The aim is to stack the cubes on top of each other into a 4 × 1 × 1 cuboid so that there are four different colours on each face of the cuboid.

Blue	Green	Red	Yellow	
Red	Blue	Yellow	Green	Blue

The cubes are painted as follows.

CUBE 1	CUBE 2	CUBE 3	CUBE 4
Blue opposite Green	B opp G	R opp B	Y opp R
Red opposite Red	G opp R	R opp Y	Y opp G
Yellow opposite Blue	B opp Y	G opp Y	B opp G

Draw a graph with four vertices, one for each colour. Blue and green are to be connected by an edge labelled 1, because cube 1 has a pair of opposite faces coloured blue and green. There are 11 more edges to draw similarly. You can solve the problem by splitting your graph into sub-graphs, each containing edges labelled 1, 2, 3 and 4, and with each vertex of order 2. The first sub-graph shows which colours are arranged at the front/back of each cube and the second which are arranged at the top/bottom. The third gives the side to side arrangement, which is how they are placed together. (There are three possible solutions.)

11* Vertices in a network are said to be adjacent if there is an edge joining them.

Colours c1, c2, c3, ... are to be assigned to the vertices (a, b, c, ...) of a graph so that no two adjacent vertices share the same colour. A greedy algorithm for achieving this is as follows:

1　Allocate colour c1 to vertex a.
2　Choose the next uncoloured vertex in alphabetical order. List the colours of all the vertices adjacent to it that are already coloured. Choose the first colour that is not in that list and allocate that colour to the vertex.
3　Repeat step 2 until all vertices are coloured.

(i) **(a)** Use the algorithm to complete the colouring of the following network, taking the vertices in alphabetical order.

Indicate the colours that you allocate (c1, c2, etc.) in the boxes provided.

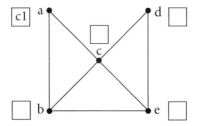

(b) Give a colouring which uses fewer colours.

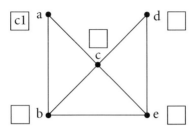

(ii) (a) A loop is an edge which connects a vertex to itself. Explain why it is not possible to colour vertices in networks with loops.

(b) Explain why repeated edges can be ignored when colouring vertices in a network.

(iii) At a mathematics open day, 6 one-hour lectures a, b, c, d, e and f are to be scheduled so as to allow students to choose from the the following combinations:

a & b;   a & d;   c & e;   b & f;   d & e;   e & f;   a & f.

(a) Represent this as a network, connecting two vertices with an edge if they must **not** be scheduled at the same time.

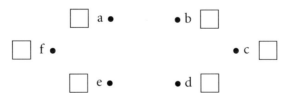

(b) Use the algorithm to colour the network, taking the vertices in alphabetical order. Show the colours that you allocate (c1, c2, etc.) on your network.

(c) Say how the colouring can be used to give a schedule for the lectures, and give the schedule implied by your colouring.

(d) Find a best schedule for the lectures, i.e. one which is completed in the shortest time possible.

[MEI]

Extension work.

*Questions marked * may be suitable starting points for coursework.*

1 Show that the following two graphs are isomorphic.

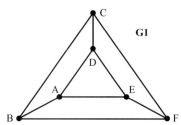 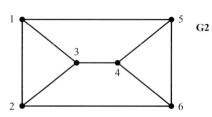

Draw a simple connected graph on six vertices, each of degree 3, which is not isomorphic to G1/G2.

2 **(i)** Find a Hamiltonian cycle for this graph.

(ii) Starting with your cycle show how to construct a diagram with no arcs crossing to show that the graph is planar.

[AEB, adapted]

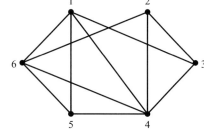

3 The travelling sales rep problem is to find the shortest Hamiltonian cycle in a network, i.e. the shortest path visiting each vertex *once and only once* and returning to the initial vertex.

(i) Regarding ABCDEA as different from AEDCBA, how many different Hamiltonian cycles are there in the complete graph on five vertices, as shown in the diagram?

(ii) How many Hamiltonian cycles are there in the complete graph on fifty vertices?

A fast computer takes 1 second to find the lengths of 10 million such cycles.

Approximately how many years will it take to find the lengths of all of them?

[AEB, adapted]

4 **(i)** A, B, C and D are the vertices of the complete graph, K_4. List all the paths from A to B (i.e. routes passing through particular vertices at most once).

(ii) How many paths are there from A to B in the complete graph on the vertices {A, B, C, D, E}?

[AEB]

5 There are six people in a room. Prove that there must be at least three people who know each other, or at least three people none of whom know one another.

6 There are a number of people at a party, and some handshaking has taken place. Prove that there must be two people who have shaken hands with the same number of people. (You will need a bipartite graph for this. One of the sets has a vertex for each person. The other set has vertices representing numbers.)

7* This question concerns a measuring problem of a type which many people find interesting and challenging. Graph theory does not provide a mechanism for solving it, but it does provide a helpful way of keeping track of what has been tried.

The problem: Two full containers each hold 10 litres of a liquid. Two empty containers are available with capacities 5 litres and 4 litres. Without using any other container, and without spilling or losing any liquid, find out how to obtain 3 litres in each of the currently empty containers.

The graph: The start vertex is (10, 10, 0, 0). From the vertex a move can be made to (10, 5, 5, 0) or (10, 6, 0, 4).

(It might be argued that other possibilities are (5, 10, 5, 0) and (6, 10, 0, 4) but these do not add anything – so let us agree to never having the first number less than the second.)

Continue until a vertex of the form (*, *, 3, 3) is obtained.

(In problems like this it can sometimes help to work backwards from the target!)

KEY POINTS

1 A *graph* consists of vertices and edges.

2 A *connected graph* has no separate parts. A *simple graph* has no loops and no more than one edge between any pair of vertices.

3 The *degree* (or *order*) of a vertex is the number of edges incident on it. A *Eulerian graph* is a connected graph in which all of the vertices are of even order.

4 A *cycle* is a closed path; the end of one edge is the start of the next and no vertex is repeated.

5 A *tree* is a simple connected graph containing no cycles.

6 A *digraph* is a graph containing at least one directed edge.

7 A *complete graph* is a simple graph in which every pair of vertices is connected by an edge.

8 An *incidence* matrix is a way of representing a graph by a matrix.

9 A *planar graph* is one which can be drawn without any edges crossing.

10 A *biparatite graph* is one in which the vertices fall into two sets and in which each edge has a vertex from one set at one end and from the other set at its other end.

3 Networks

'Tis true; there's magic in the web of it.

Shakespeare, Othello

A network is a *weighted graph* – that is a graph in which there is a number (weight) associated with each edge. An example is shown in figure 3.1.

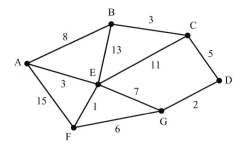

Figure 3.1

Obvious examples of networks include maps and similar geographical networks. Here the vertices represent geographical points. For a map the edge weights are distances, but other geographical networks might have weights representing costs or times.

Networks have a much wider applicability than just geographical situations, but the two problems covered in this chapter, the minimum connector and the shortest path, are easiest to visualise in a geographic context.

The minimum connector problem

A cable TV company based in Plymouth wishes to make connections in the most economical way to all the towns in the south west shown on the map (see figure 3.2). They must therefore link all towns using as little cable as possible.

❓ Try to solve this problem before you read any further. Later on you can compare the strategy you used with the algorithms given in the chapter.

Consider the network in figure 3.3(a) with its arcs removed (as in figure 3.3(b)). The minimum connector problem is to make a selection of the available arcs so

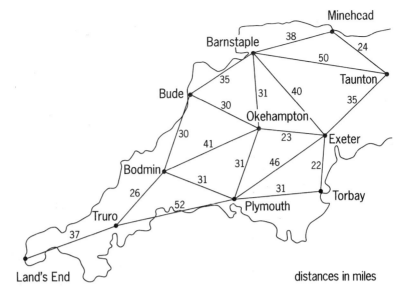

Figure 3.2

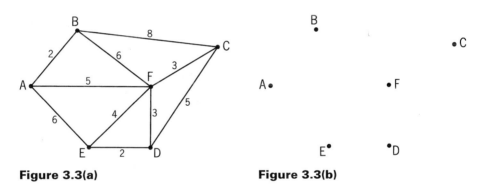

Figure 3.3(a)   Figure 3.3(b)

that any one node can be reached from any other, and such that the total length of the chosen arcs is as small as possible. Clearly the resulting choice will contain no loops or cycles: if A, B and F are linked as shown in figure 3.4, B is already connected to F via A so it is unnecessary to choose arc BF.

A connected set of arcs with no loops is called a *tree*, and the set which solves the minimum connector problem is referred to as the *minimal spanning tree* for the network. The problem of identifying the minimal spanning tree is a real one for cable TV companies, since provided each town is connected in some way to the base station it will still receive the signal. Thus the only concern of

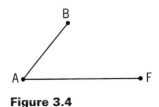

Figure 3.4

the company is to keep the total length of cable required to a minimum. For this reason, the problem is often called the cable TV problem.

Minimal spanning tree algorithms

You will consider two similar algorithms that can be used to solve the cable TV problem. Both are examples of 'greedy algorithms': ones that maximise immediate rewards regardless of future consequences. Although such methods do not always lead to optimal solutions, the minimum connector algorithms that follow produce optimal solutions. However, one of them is more amenable to computer implementation than the other.

Kruskal's algorithm

First select the shortest arc of the network.

At each subsequent stage, select from those arcs that have not yet been selected the shortest arc that does not link nodes between which a path has already been created. For a network with n nodes, when $n - 1$ arcs have been selected a minimal spanning tree will have been found.

If at any stage there is a choice of shortest arcs, choose arbitrarily between them.

Let us look at the stages involved in applying the algorithm to the network in figure 3.5.

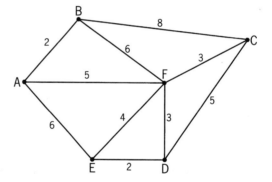

Figure 3.5

Arcs ranked in order of increasing length.

Length	Arcs
2	AB, DE
3	CF, DF
4	EF
5	CD, AF
6	AE
8	BC

1 We can start by selecting AB or DE, so let us arbitrarily select AB as shown in figure 3.6.

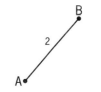

Figure 3.6

2 Select DE as in figure 3.7.

Figure 3.7

3 We can select CF or DF, so let us arbitrarily select DF as in figure 3.8.

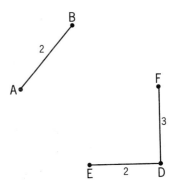

Figure 3.8

4 Select CF (figure 3.9).

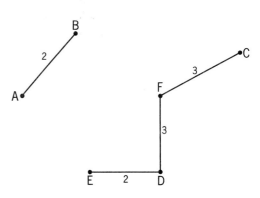

Figure 3.9

5 The next shortest arc is EF but E and F are already connected via D, so we do not select EF.

6 The next shortest arcs are CD and AF but C and D are already connected via F, so we choose AF (figure 3.10).

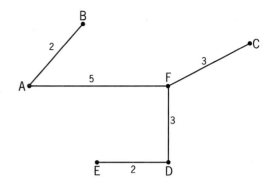

Figure 3.10

Five arcs have now been selected and so all six nodes are connected. The minimal spanning tree has a length of 15.

Prim's algorithm

First select an arbitrary node, then connect it to the nearest node.
Now connect the nearest node that is not already connected, to those already in the solution.

Repeat this until all nodes have been connected.

Here is Prim's algorithm applied to the network we used to demonstrate Kruskal's algorithm.

1 Select an arbitrary node, F say. Connect it to the nearest node: C and D are both distance 3 away, so arbitrarily choose C as in figure 3.11.

Figure 3.11

2 D is the nearest node not yet in the solution so we connect this next (figure 3.12).

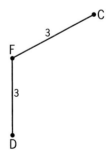

Figure 3.12

3 E is now the nearest node not yet in the solution so this is connected next (figure 3.13).

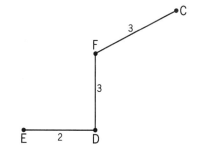

Figure 3.13

4 A is now the nearest node not yet in the solution so this is now connected (figure 3.14).

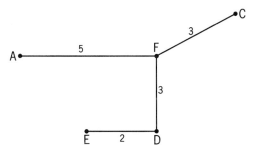

Figure 3.14

5 Finally connect B. Notice that this gives the same minimal spanning tree as before, of length 15 (figure 3.15).

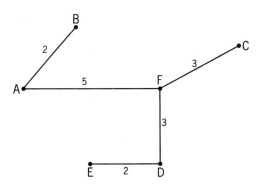

Figure 3.15

Find the minimal spanning tree and associated shortest distance for each of the networks below.

1

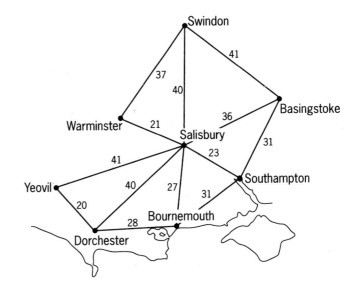

2

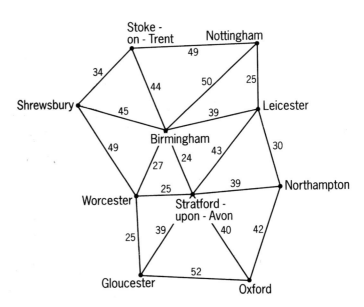

3

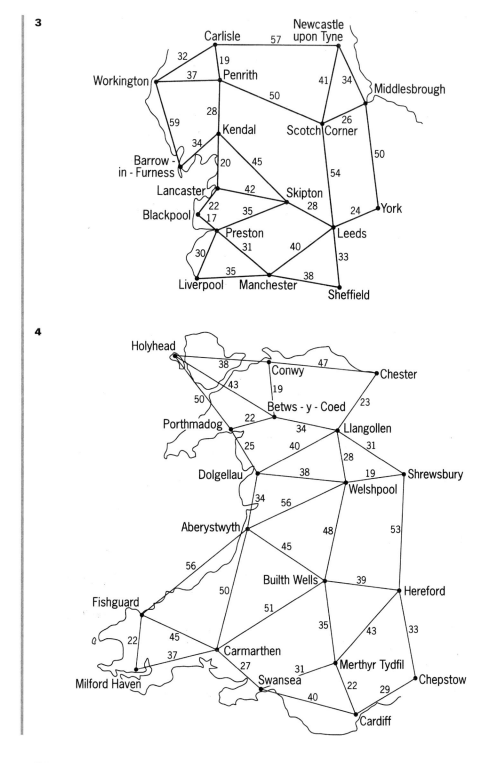

4

Distance tables and networks

Now that you have had the opportunity to apply the algorithms to some networks, you can consider how to apply them when the network is given in the form of a distance table. This would be the form in which you would need to supply the network details to a computer, so in this form the algorithms could be automated.

Kruskal's algorithm is not particularly well suited to computer implementation as it requires the arcs to be arranged in order of length. This would mean using one of the sorting algorithms from Chapter 1. It also needs a way of recognising when cycles might be created, which is particularly tricky on a computer. You might like to consider why it is difficult to write such an algorithm by having a go at writing one yourself.

Prim's algorithm, on the other hand, is far more suitable for computerisation. The distance table for the cable TV problem discussed earlier is shown below.

	A	B	C	D	E	F
A	–	2	∞	∞	6	5
B	2	–	8	∞	∞	6
C	∞	8	–	5	∞	3
D	∞	∞	5	–	2	3
E	6	∞	∞	2	–	4
F	5	6	3	3	4	–

(∞ indicates no direct link)

1 Select an arbitrary node, say F, and delete its row. (This is equivalent to ignoring nodes that are already connected in the network method.) Look for the smallest entry in the column for the selected node, in this case F. There are two 3s so arbitrarily choose the one in row C. Add the new node, in this case node C, to the solution with arc CF. This is shown in figure 3.16 so that you can compare the method with diagrammatic approaches.

	A	B	C	D	E	F
A	–	2	∞	∞	6	5
B	2	–	8	∞	∞	6
C	∞	8	–	5	∞	3
D	∞	∞	5	–	2	3
E	6	∞	∞	2	–	4

Figure 3.16

2 Delete row C and look for the smallest entry in columns F and C (figure 3.17).

The 3 in column F, row D, is the smallest so we add node D to the solution with arc DF.

	A	B	C	D	E	F
A	–	2	∞	∞	6	5
B	2	–	8	∞	∞	6
D	∞	∞	5	–	2	3
E	6	∞	∞	2	–	4

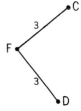

Figure 3.17

3 Delete row D and look for the smallest entry in columns F, C and D. The 2 in column D, row E, is the smallest so we add node E to the solution with arc ED (figure 3.18).

	A	B	C	D	E	F
A	–	2	∞	∞	6	5
B	2	–	8	∞	∞	6
E	6	∞	∞	2	–	4

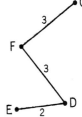

Figure 3.18

4 Delete row E and look for the smallest entry in columns F, C, D and E.

The 5 in column F, row A, is the smallest, so we add node A to the solution with arc FA (figure 3.19).

	A	B	C	D	E	F
A	–	2	∞	∞	6	5
B	2	–	8	∞	∞	6

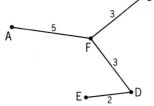

Figure 3.19

5 Finally, delete row A, and look for the smallest entry in columns F, C, D, E and A. At this stage B is the only row left, because B is the only node not connected.

The smallest entry is the 2 in column A, row B, so we add node B to the solution with arc AB (see figure 3.20).

	A	B	C	D	E	F
B	2	–	8	∞	∞	6

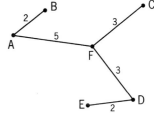

Figure 3.20

Note

In practice, you would not have to rewrite the table at each stage but simply circle the chosen entry, cross out the rows of the original table, and progressively tick the columns from which you can select.

1 Find the minimal spanning tree for the following network which is given in tabular form.

	Malvern	Worcester	Hereford	Evesham	Ross	Tewkesbury	Gloucester	Cheltenham
Malvern	–	8	19	∞	19	13	20	∞
Worcester	8	–	25	16	∞	15	∞	∞
Hereford	19	25	–	∞	14	∞	28	∞
Evesham	∞	16	∞	–	∞	13	∞	16
Ross	19	∞	14	∞	–	24	16	∞
Tewkesbury	13	15	∞	13	24	–	10	9
Gloucester	20	∞	28	∞	16	10	–	9
Cheltenham	∞	∞	∞	16	∞	9	9	–

2 Find the minimal spanning tree for this network.

	Dorchester	Puddletown	Blandford	Wimborne	Bere Regis	Lytchett Minster	Weymouth	Warmwell	Wareham	Swanage	Poole
Dorchester	–	5	∞	∞	∞	∞	8	5	∞	∞	∞
Puddletown	5	–	12	∞	6	∞	∞	9	14	∞	∞
Blandford	∞	12	–	7	9	11	∞	∞	16	∞	∞
Wimborne	∞	∞	7	–	8	7	∞	∞	∞	∞	7
Bere Regis	∞	6	9	8	–	8	19	11	8	∞	∞
Lytchett Minster	∞	∞	11	7	8	–	25	∞	5	∞	6
Weymouth	8	∞	∞	∞	19	25	–	7	∞	∞	∞
Warmwell	5	9	∞	∞	11	∞	7	–	13	∞	∞
Wareham	∞	14	16	∞	8	5	∞	13	–	10	∞
Swanage	∞	∞	∞	∞	∞	∞	∞	∞	10	–	∞
Poole	∞	∞	∞	7	∞	6	∞	∞	∞	∞	–

Problems 1 and 2 introduce situations in which there are only nodes, no established arcs. This gives you a free choice in selecting arcs. To produce a minimal spanning tree we would draw arcs from node to node, and would not create any new nodes, but you might like to consider whether a relaxation of this requirement would lead to a better solution to these two problems.

1 Figure 3.21 shows a set of points on a printed circuit board which need to be joined together as economically as possible (i.e. using the shortest possible length of conductor), so that they can all be supplied with power. How should this be done?

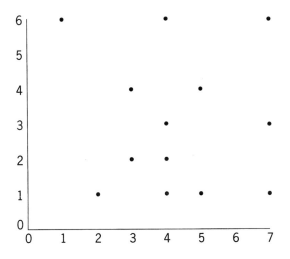

Figure 3.21

2 The UKG gas company has six sites in the North Sea. The locations are shown on figure 3.22. Pipelines are to be laid to take the oil from the fields to Aberdoon on shore. Advise on the most economic way to achieve this.

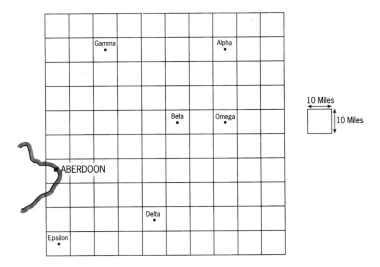

Figure 3.22

3 In practice, networks are not usually as simple as the diagrams we have been working on would suggest, and some preparatory work must be done to extract the information relevant to the problem in hand. Simplify the map shown in figure 3.23 and hence find the minimum rail network that will still allow rail connections between the towns shown.

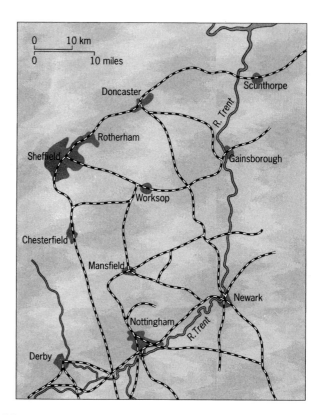

Figure 3.23

4 A cable TV company based in Hull wishes to connect the towns shown in the map (figure 3.24). They wish to know whether or not it is worth approaching the owners of the Humber Bridge with a view to running cables over the bridge. Minimum inter-town distances, with and without making use of the bridge, are given on the following pages in kilometres. Investigate the saving in cable that would result from using the bridge. (Answer provided.)

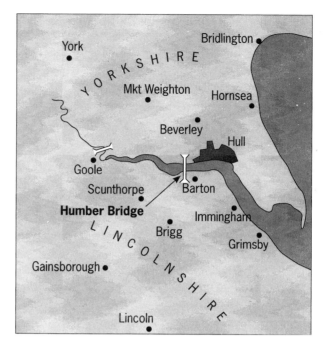

Figure 3.24

Table 1: Without using bridge.

	1	2	3	4	5	6	7	8	9	10	11	12	13	14
1 Hull		12	60	51	90	77	97	35	126	93	121	119	31	42
2 Beverley			47	32	87	74	103	23	122	90	122	116	18	39
3 York				64	87	74	103	69	122	90	122	116	29	39
4 Bridlington					119	106	135	26	154	122	154	148	48	71
5 Brigg						12	16	103	39	31	35	29	56	48
6 Scunthorpe							29	109	48	27	48	42	58	35
7 Barton								122	35	47	32	26	87	64
8 Hornsea									149	113	142	138	40	61
9 Lincoln										29	55	55	106	84
10 Gainsborough											56	56	74	51
11 Grimsby												16	106	84
12 Immingham													100	77
13 Mkt Weighton														23
14 Goole														

Table 2: Using the bridge.

	1	2	3	4	5	6	7	8	9	10	11	12	13	14
1 Hull		12	60	51	26	42	13	35	69	56	53	45	31	42
2 Beverley			47	32	64	55	26	23	74	69	66	58	18	39
3 York				64	73	74	73	69	122	87	113	105	18	39
4 Bridlington					73	93	64	26	113	108	105	97	48	71
5 Brigg						12	16	87	39	31	35	29	39	48
6 Scunthorpe							29	77	48	27	48	42	58	35
7 Barton								48	55	47	32	26	43	39
8 Hornsea									97	92	89	80	40	61
9 Lincoln										29	55	55	98	84
10 Gainsborough											56	56	74	51
11 Grimsby												16	84	68
12 Immingham													77	61
13 Mkt Weighton														23
14 Goole														

5 Write a computer program to apply Prim's algorithm to a network presented in tabular form.

6 (i) Four new villages are to be built at the corners of a square of side 10 km (see figure 3.25). The local authority is very short of money and wishes to design a road system to link them together as economically as possible. The first plan that was presented to them was simply a square of four roads with a total distance of 40 km. The treasurer said this was out of the question. 'We can't even afford 30 km' he told the Planning Committee. Can you help?

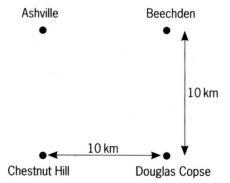

Figure 3.25

(ii) The inhabitants of the planet Morg live underground. They are excavating a new structure consisting of eight spherical chambers at the corners of a cube of side 1 km. They wish to link the chambers by tunnels. What is the most economic way to achieve this, assuming that tunnelling is equally expensive in all directions?

(Answers provided.)

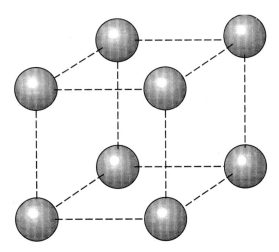

Figure 3.26

Finding a shortest path

Autoroute is a computerised route planner. You provide your starting point and destination and the computer prints out the best route together with directions, as you can see in figure 3.27.

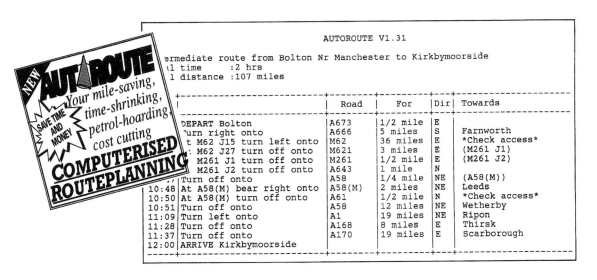

```
                              AUTOROUTE V1.31

ermediate route from Bolton Nr Manchester to Kirkbymoorside
l time     :2 hrs
l distance :107 miles
```

	Road	For	Dir	Towards
DEPART Bolton	A673	1/2 mile	E	
Turn right onto	A666	5 miles	S	Farnworth
t M62 J15 turn left onto	M62	36 miles	E	*Check access*
M62 J27 turn off onto	M621	3 miles	E	(M261 J1)
M261 J1 turn off onto	M261	1/2 mile	E	(M261 J2)
M261 J2 turn off onto	A643	1 mile	N	
Turn off onto	A58	1/4 mile	NE	(A58(M))
10:48 At A58(M) bear right onto	A58(M)	2 miles	NE	Leeds
10:50 At A58(M) turn off onto	A61	1/2 mile	N	*Check access*
10:51 Turn off onto	A58	12 miles	NE	Wetherby
11:09 Turn left onto	A1	19 miles	NE	Ripon
11:28 Turn off onto	A168	8 miles	E	Thirsk
11:37 Turn off onto	A170	19 miles	E	Scarborough
12:00 ARRIVE Kirkbymoorside				

Figure 3.27

 What method does such a program use?

Another technological innovation to help the motorist negotiate the traffic in a busy city is the 'in-car navigation system'. The motorist is in contact with a central computer which is constantly monitoring the state of traffic flow through a network of sensors in the city streets. The motorist simply keys his or her location and chosen destination into a small communication device located on the dashboard. The computer decides the best route based on the latest traffic information and transmits the necessary directions back to the driver at the appropriate moments. It keeps track of the car's progress using its network of sensors.

As with *Autoroute*, at the heart of this system is an algorithm to find the shortest route between two points. Before considering such an algorithm, try the following exercise based on a network representing the centre of a small town.

EXERCISE 3C

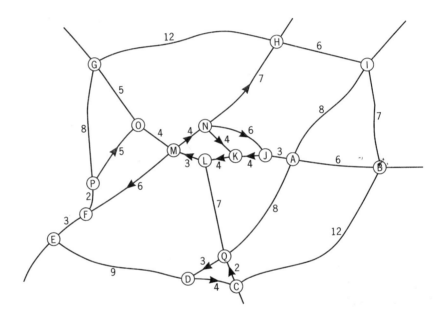

Junctions are represented by points or *nodes* and the roads by lines or *arcs*. The numbers on the arcs are the times in minutes that it takes to travel between the junctions. One-way streets are marked by arrows.

1 What are the quickest routes between the following places?

> **(i)** From F to A   **(ii)** from B to G
> **(iii)** from E to A   **(iv)** from H to Q
> **(v)** from N to C

2 (i) The fire station is located at the point C. Suppose there is a call to attend a fire at point G, what route should the fire engine take?
> **(ii)** What would be the quickest way back to the station?

3 There has been a burglary at a shop at L. Police cars are located at points F, G and H when the alarm is raised. Which one can get to the scene of the crime first?

4 Suppose it is decided that the road NK should be a traffic-free shopping precinct. What effect will this have on the quickest route from G to L?

5 What is the best route from the hospital at O to the point I where there has been a road accident?

6 What is the quickest route from J to N?
Suppose there has been an accident in road KL blocking it. How long a delay would make it worth my while finding an alternative route?

7 The fire station is at C. Consider how you might approach the problem of deciding if it would be better at F.

Developing an algorithm for a shortest route

In the network shown in figure 3.28 the numbers represent the lengths of the arcs. Find the shortest route from S to T.

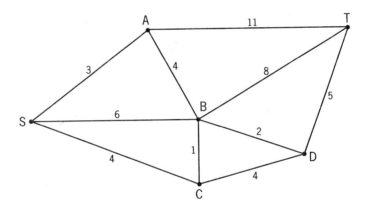

Figure 3.28

The solution to this problem can easily be found by inspection, but we want to develop a systematic approach that can be automated to apply to larger problems. Here is a procedure.

Label the start node S with 0.

Consider those nodes that can be reached directly from S, in this case A, B and C. Put temporary labels on A, B, C equal to their direct distances from S, in this case 3, 6 and 4 respectively.

Select the node with the smallest temporary label (in this case A) and make its label permanent by putting a box around it. This indicates that the shortest distance from S to A is 3, and that it cannot be improved upon. Strictly the label on S is permanent and should have a box around it too, so at this stage we have the situation shown in figure 3.29.

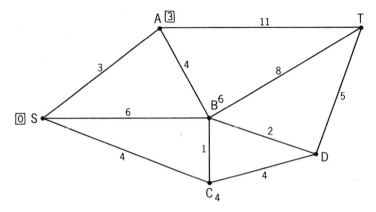

Figure 3.29

Consider all nodes that can be reached directly from A, in this case B and T.

The shortest route from S to B via A is $3 + 4 = 7$ (using the fact that we know the shortest route from S to A is 3). But B is already labelled with 6 (direct route from A) so we retain the 6 as the best so far.

The shortest route from S to T via A is $3 + 11 = 14$ so we put a temporary label of 14 on T.

The minimum temporary label is now 4 at node C. Make this permanent to indicate that the shortest route from S to C is 4 and that it cannot be improved upon. The situation is shown in figure 3.30.

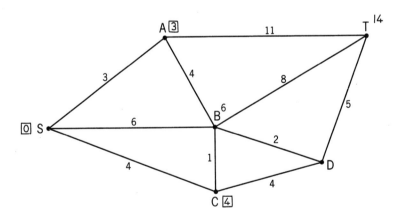

Figure 3.30

Consider all nodes that can be reached directly from C, in this case B and D.

The shortest route from S to B via C is $4 + 1 = 5$, which is shorter than the present label so we change the temporary label of B to 5. (It is best to show this change by crossing out the 6, but leaving it visible.)

The shortest route from S to D via C is $4 + 4 = 8$ so we put a temporary label of 8 on D.

The minimum temporary label is now 5 at node B. Make this permanent to indicate that the shortest route from S to B is 5 and that it cannot be improved upon. Figure 3.31 shows the stage that we have now reached.

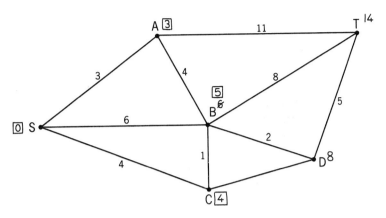

Figure 3.31

Consider all nodes directly connected from B, i.e. D and T.
The shortest route from S to D via B is 5 + 2 = 7, which is shorter than that on the present label, so we change it to 7.

The shortest route from S to T via B is 5 + 8 = 13, shorter than that on the present label so we change it to 13.

The minimum temporary label is now 7 at node D. Make this permanent to show that the shortest route from S to D is 7. Figure 3.32 shows the new situation.

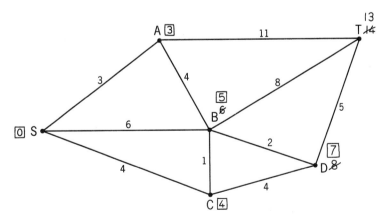

Figure 3.32

Consider all nodes that can be reached directly from D, i.e. just T.

Replace the label at T with 7 + 5 = 12 since this is less than its present 13.

The minimum temporary label at node T is now 12. Make this permanent.

The destination node now has a permanent label as shown in figure 3.33, so we now know that the shortest route from S to T is 12. It remains only to find the route.

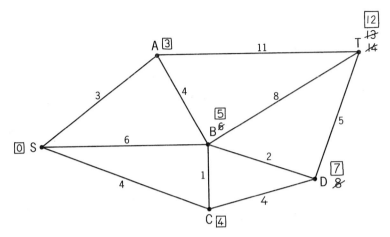

Figure 3.33

To find the shortest route we work backwards from T. The final stage brought us via D. If we look back through the working we will find that the best route to D was via B. It would have been a good idea to add to our labels the letter of the node from which we had found the best route. This would have saved us the trouble of looking back through our working but would, in general, involve more work. In fact, we can simply trace back through the network from T to S along arcs for which the difference in the permanent labels is equal to the arc lengths.

Thus we arrive at the optimum route S-C-B-D-T with a distance of 12, as shown in figure 3.34.

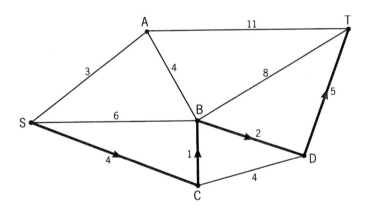

Figure 3.34

Dijkstra's algorithm for a shortest route

The algorithm that we have built up here was actually first described by Dijkstra, and has become known as *Dijkstra's algorithm for the shortest route*. The steps in the algorithm are written in general form as follows.

STEP 1

Label start node S with permanent label (P-label) of 0.

For all nodes that can be reached directly from S, assign temporary labels (T-labels) equal to their direct distance from S.

Select the node with the smallest T-label and make its label permanent.

The P-label represents the shortest distance from S to that node.

STEP 2

Put a T-label on each node that can be reached directly from the node that has just received a P-label. The T-label must be equal to the sum of the P-label and the direct distance from it. If there is an existing T-label at a node, it should be replaced only if the new sum is smaller.

Select the minimum T-label and make it permanent.

If this labels the destination node go to step 3, otherwise repeat step 2.

STEP 3

To find the shortest path(s), trace back from the destination including any arc MN for which

(P-label of M) – (P-label of N) = length of arc MN.

EXERCISE 3D

1 Use Dijkstra's algorithm to find a shortest path from S to T for the networks (i), (ii) and (iii) below.

(i)

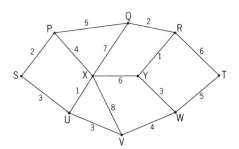

(ii)

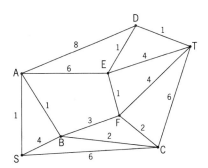

(iii)

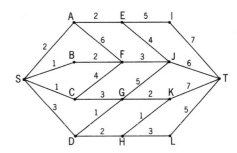

2 The map shows the main railway lines across the USA and gives the approximate times in hours for the various journeys.

Find the quickest route

(i) from Los Angeles to Chicago

(ii) from New Orleans to Denver.

(iii) If you can travel by road from El Paso to Santa Fe in 5 hours and from Santa Fe to Denver in 5 hours, would you save time on journey (i) or (ii) by using a mix of road and rail? (You should neglect connection times.)

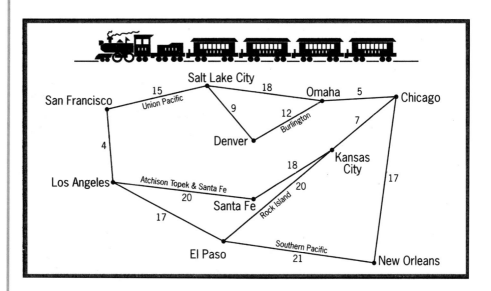

3 The fire department in Westingham has a team fighting a large blaze at one of the town's hotels. They urgently need extra help from one of the neighbouring towns, A, B or C. The estimated times (in minutes) to travel along the various sections of road from A, B and C to Westingham are shown on the network on the next page. Which town's fire fighters should they call upon and how long will it take them to arrive at the fire?

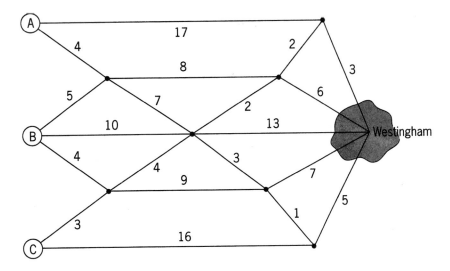

4 A company in the West Midlands has two factories, one in Kidderminster and the other in Cheltenham. A van has to travel regularly between the two factories. Use the map of the M5 between junctions 4 and 10 overleaf to answer the questions below.

(The numbers on the arcs are the distances in miles and the numbers in the circles are the motorway junction numbers.)

(i) For the journey from Kidderminster to Cheltenham

(a) What is the shortest route?
(b) If we assume an average speed on the motorway of 60 mph and on other roads of 40 mph, what is the quickest route?
(c) What speed would you need to average on the motorway to make it worth joining the M5 at an earlier junction than in (b)?
(d) Roadworks are scheduled on the M5 between junctions 6 and 7, reducing the average speed over this section to 20 mph. Will this affect the route in (b)? If so, where should the motorway now be joined?

(ii) The managing director of the company lives in Tewkesbury. Using the speeds given in (i) (b), what routes should he use to travel to the factories in

(a) Kidderminster
(b) Cheltenham?
(c) Will his best route to Kidderminster be affected by the forthcoming roadworks?

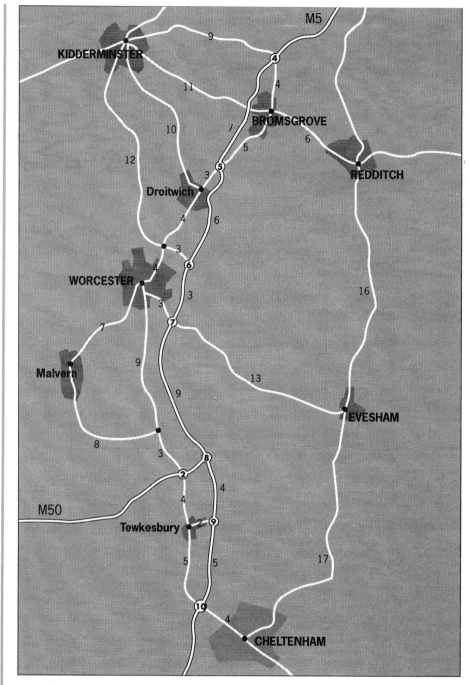

Map to show the M5 between junctions 4 and 10

1 You have only applied Dijkstra's algorithm to a network diagram. Consider how you would adapt the algorithm if the network information were presented in the form of a distance table like this.

	Dorchester	Puddletown	Blandford	Wimborne	Bere Regis	Lytchett Minster	Weymouth	Warmwell	Wareham	Swanage	Poole
Dorchester	–	5	∞	∞	∞	∞	8	5	∞	∞	∞
Puddletown	5	–	12	∞	6	∞	∞	9	14	∞	∞
Blandford	∞	12	–	7	9	11	∞	∞	16	∞	∞
Wimborne	∞	∞	7	–	8	7	∞	∞	∞	∞	7
Bere Regis	∞	6	9	8	–	8	19	11	8	∞	∞
Lytchett Minster	∞	∞	11	7	8	–	25	∞	5	∞	6
Weymouth	8	∞	∞	∞	19	25	–	7	∞	∞	∞
Warmwell	5	9	∞	∞	11	∞	7	–	13	∞	∞
Wareham	∞	14	16	∞	8	5	∞	13	–	10	∞
Swanage	∞	∞	∞	∞	∞	∞	∞	∞	10	–	∞
Poole	∞	∞	∞	7	∞	6	∞	∞	∞	∞	–

∞ means no direct connection.

2 The following is an alternative algorithm, due to Ford, for finding the shortest distance between two nodes of a network. Investigate how it works and compare it with Dijkstra's algorithm.

Step 1   Let X_0 be the departure node. Put the value 0 against X_0 and values of ∞ against the other nodes.

Step 2   Apply the following rule: if λ_i is the value against X_i, look for an arc (X_i, X_j) such that
$\lambda_j - \lambda_i > d(X_i, X_j)$
where $d(X_i, X_j)$ is the length of arc (X_i, X_j).
Replace λ_j by $\lambda_i + d(X_i, X_j)$.
Continue until there is no arc which will lead to a decrease in any λ_j. The resulting values of λ_j will then be the shortest distances from X_0 to each node X_j.

Step 3   The shortest path can then be found by the same procedure as in the Dijkstra algorithm.

3 Solve the pair of problems below and consider how you could adapt Dijkstra's algorithm to solve such problems in general. Note that in these networks the nodes are clearly marked and that arcs do not meet at the other crossings. These would be places where the routes went over or under each other.

(i) Weight restriction problem

The numbers on the arcs in the network below are the maximum weights in tons that those arcs can withstand (because of bridges with weight restrictions and so on). What is the heaviest vehicle that can be driven from S to T?

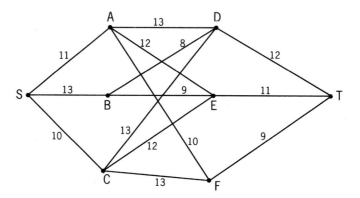

Figure 3.35

(ii) Freight aircraft problem

A freight aircraft wishes to fly from S to T. In order to maximise its cargo load it needs to keep the fuel carried to a minimum. This means that short hops are preferable to a long haul. You require the route for which the longest hop is as small as possible. What route should you choose?

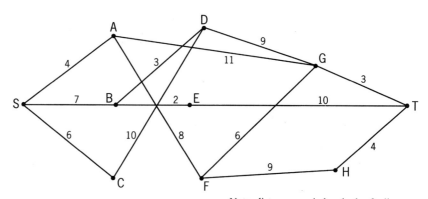

Note: distances are in hundreds of miles

Figure 3.36

(Answer provided.)

4 A haulage contractor has a lorry at town S. She needs to get it to town T at minimum cost. The network in figure 3.37 represents the roads connecting S and T. The numbers alongside each edge give the costs in £s of sending the lorry along that road. In one case the cost in one direction along a road is positive, but in the other is negative, representing profit, because the lorry may collect and deliver a number of parcels *en route*, if the contractor so wishes.

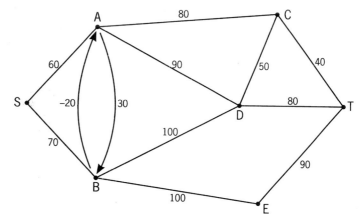

Figure 3.37

(i) Explain why Dijkstra's algorithm fails in this example and suggest how this problem might be overcome.

(ii) Test your ideas on the problem shown in figure 3.38, in which the aim is again to travel from S to T at the minimum cost.

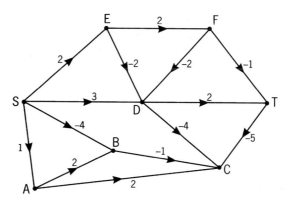

Figure 3.38

(Answer provided.) [JMB]

5 Adapt Dijkstra's algorithm to find the *longest* path through a directed network which contains no cycles.

6 Figure 3.39 is a map of a small section of the London Underground. Investigate the quickest route between pairs of named stations. Assume that the time to travel one section (between neighbouring stations) is 2 minutes and that transfer between lines including waiting time takes 6 minutes.

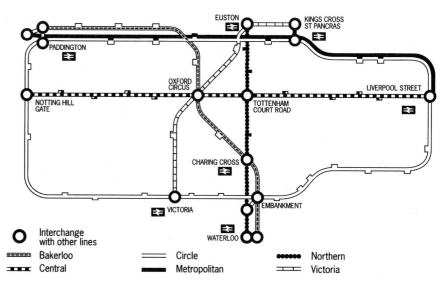

Figure 3.9

7 Write a computer program for Dijkstra's algorithm.

Complexities

For each of the three algorithms (Kruskal, Prim and Dijkstra), the work involved in applying the algorithm is related to the number of vertices in the network. The number of edges matters as well, but complexity is defined by the worst possible case (see 'Algorithmic complexity' in Chapter 1). So far, for each algorithm you need to relate the work involved to the number of vertices for a *complete* network, i.e. a network in which each vertex is connected to every other vertex. For example, see figure 3.40.

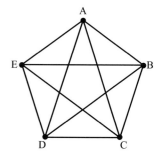

Figure 3.40

Kruskal

The first step in Kruskal is to choose the edge with the smallest weight. The work involved in making this choice will be proportional to the number of edges less 1, because that is the number of comparisons which will have to be made. There are $\frac{n(n-1)}{2}$ edges in a complete network on n vertices, so the work involved in this will be proportional to $\frac{n(n-1)}{2} - 1$.

Note

In the complete network on 5 vertices drawn in figure 3.40 there are $\frac{5 \times 4}{2} = 10$ edges.

In the second step the smallest of the remaining $\frac{n(n-1)}{2} - 1$ edges is chosen. In total $(n-1)$ edges are chosen, so the work involved is proportional to

$$\left(\frac{n(n-1)}{2} - 1\right) + \left(\frac{n(n-1)}{2} - 2\right) + \left(\frac{n(n-1)}{2} - 3\right) + \ldots + \left(\frac{n(n-1)}{2} - (n-1)\right)$$

You need not worry about the detail of the algebra but the expression can be written as $\frac{1}{2}n(n-1)(n-2) = \frac{1}{2}(n^3 - 3n^2 + 2n)$, showing that the algorithm has cubic complexity.

❓ Check that the cubic expression gives the correct value when n = 5 (i.e. $9 + 8 + 7 + 6 = 30$).

Note

No account has been taken of the extra work that is involved at each step in checking that no connection has yet been established between the vertices which are to be connected. The work involved is not sufficient to affect the complexity.

Prim

In step 1 the smallest weight edge out of $n - 1$ possibilities has to be chosen.

In step 2 the connected set contains 2 vertices and there are $n - 2$ possible unconnected vertices to consider. Thus the least weight edge out of $2(n - 2)$ edges must be chosen. Continuing with this argument the work involved will be proportional to

$$((n-1) - 1) + (2(n-2) - 1) + (3(n-3) - 1) + \ldots + ((n-1) - 1).$$

The algebra of this is difficult but you can see that the general term is quadratic $(i(n - i) - 1)$, and that there are $n - 1$ terms (i runs from 1 to $n - 1$). So the expression's highest order term will be a term in n^3.

Thus both minimum connector algorithms have cubic complexity – doubling the size of the problem will (for big problems) involve about eight times the effort.

Dijkstra

In Dijkstra the work involved at each step is approximately proportional to the number of vertices which have not been permanently labelled. So the total work involved is approximately proportional to

$$(n-1) + (n-2) + (n-3) + \ldots + 1 = \frac{1}{2}n(n-1).$$

This shows that Dijkstra has quadratic complexity. Doubling the size of the problem leads to approximately four times the effort.

1 The following matrix shows the costs of connecting together each possible pair from six computer terminals:

	A	B	C	D	E	F
A	–	120	200	140	135	250
B	120	–	230	75	130	80
C	200	230	–	160	160	120
D	140	75	160	–	200	85
E	135	130	160	200	–	150
F	250	80	120	85	150	–

The computers are to be connected together so that, for any pair of computers, there should be either a direct link between them or a link via one or more other computers. Use an appropriate algorithm to find the cheapest way of connecting these computers. Show your result on a network and give the total cost.

[AEB]

2 (i) The five points on the graph below may be connected in pairs by straight lines.

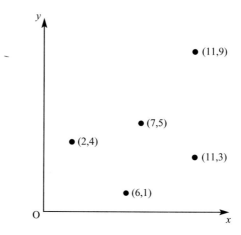

Use a greedy algorithm to find a connector of minimum length for the five points. Show your answer on a copy of the diagram, explain your method, and give the length of your minimum connector.

(ii) The points on the graph on the next page may also be connected in pairs by straight lines.
   (a) Find a minimum connector for the four points.
   (b) Find a fifth point so that the minimum connector for the five points is shorter than the minimum connector for the original four points. Show your minimum connector and give its length.

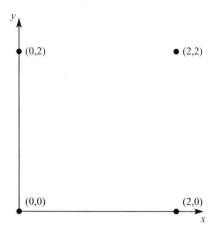

[Oxford]

3 The diagram represents the roads joining 10 villages, labelled A to J. The numbers give distances in km.

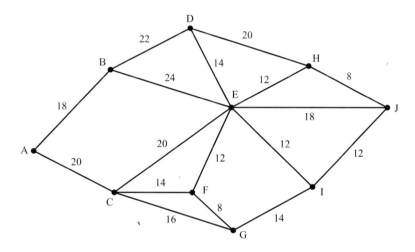

(i) Use Dijkstra's algorithm to find a shortest route from A to J. Explain the method carefully, and show all of your working. Give your shortest route and its length.

A driver usually completes this journey driving at 60 km h^{-1}. The local radio reports a serious fire at village E, and warns drivers of a delay of 10 minutes.

(ii) Describe how to modify your approach to (i) to find the quickest route, explaining how to take account of this information. What is the quickest route, and how long will it take?

[Oxford]

4 Five new houses, labelled A, B, C, D and E on the diagram overleaf, are to be connected to a drainage system, each having a connection to the sewer at the point S on the diagram, either directly or via another house. Alternatively, houses may be connected to an intermediate manhole at M, directly or via another house. This manhole must in turn be connected to S, either directly or via another house.

All connecting pipes must be such that water can drain downhill. The direction of slope is shown on the diagram.

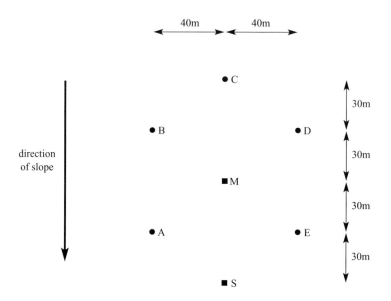

(i) Copy and complete the following matrix, showing the lengths of all 16 possible connecting pipes.

From \ To	A	B	C	D	E	M	S
A	–	–	–	–	–	–	50
B	60	–	–	–	–	50	$\sqrt{9700}$
C							
D							
E							
M							
S							

(ii) Starting from S, use Prim's algorithm to find a minimum connector for A, B, C, D, E, M and S.

Show which pipes are in your connector, indicating the order in which they were included, and give their total length.

Does your connector represent a system which drains correctly?

(iii) Investigate whether or not the provision of the intermediate manhole at M is worthwhile. Justify your conclusions.

[Oxford]

5 The network represents an electronic information network. The vertices represent computer installations. The arcs represent direct links between installations. The weights on the arcs are the rates at which the links can transfer data (in units of 100 000 bits per second).

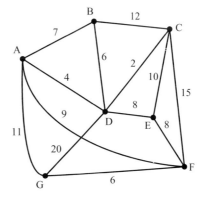

A chain of links may be established between any two installations via any other installations. For such links the rate at which the chain can transfer data is given by the speed of the slowest link in the chain.

(i) (a) The fastest chain linking G to B is needed. Devise a way of adapting Dijkstra's algorithm to produce an algorithm suited to this task. Give your adaptations and explain how your algorithm works.

(b) Apply your algorithm, showing all of your steps on a copy of the network.

(c) State the fastest chain and its speed of transferring data.

(ii) How could you find a slowest chain linking G to B?
Give such a chain.

[Oxford]

6 The diagram shows a small network in which the number on each arc represents the cost of travelling along that arc in the directions indicated.

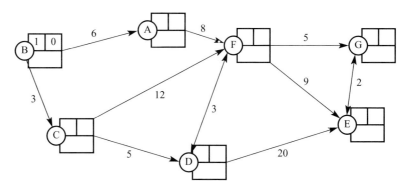

(i) Apply Dijkstra's algorithm to find the smallest cost, and the associated route, to get from B to E. (Record the order in which you permanently label nodes, and write your temporary labels and permanent labels, in boxes as shown.)

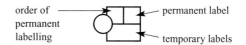

order of
permanent
labelling

permanent label

temporary labels

Show *briefly* how the route is found in Dijkstra's algorithm once the destination vertex has been permanently labelled.

(ii) Suppose now that a new route is introduced from A to C, and that profit of 4 may be made by making a delivery from A to C when using that route. This is shown as a negative cost in part of the network reproduced below.

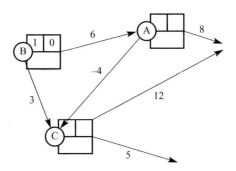

Work through Dijkstra's algorithm for this part of the network with the extra route.

Say what is the best route from B to C, and explain why Dijkstra's algorithm fails to find it.

[MEI]

7 (i) The network below consists of a set of nodes and connecting arcs. Each arc has a number (or *weight*) associated with it.

Use Dijkstra's algorithm to find the 'shortest' route from L to A in the network below, i.e. the route such that the sum of the weights on the arcs of the route is as small as possible.

(Record the order in which you permanently label nodes, and write your temporary labels and permanent labels in boxes at the vertices.)

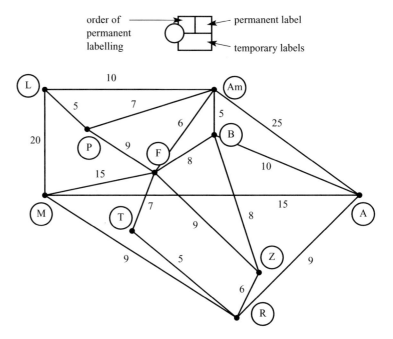

(ii) In fact the nodes of the network represent European cities, and the arcs represent air freight routes flown by a particular company. The company has a plane in London (L) which it needs to fly to Athens (A) to collect a contracted load. The weights on the arcs are the profits (£100s) which the company can make by moving loads from city to city en route. The profits are only available in the directions indicated in the diagram below. Company policy is that the plane should fly only along routes where loads are available.

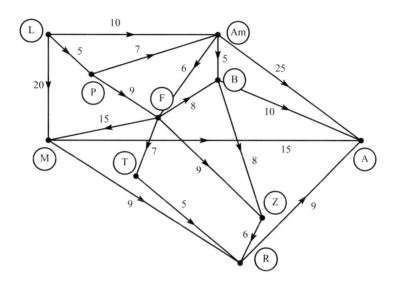

(a) Say how this problem differs from the problem in part (i).

(b) Describe how Dijkstra's algorithm could be adapted to solve this problem.

[**MEI**]

8 (i) The network overleaf represents a number of villages together with connecting roads. The numbers on the arcs represent distances in miles.

Use Dijkstra's algorithm to find the shortest routes from A to each of the other villages.

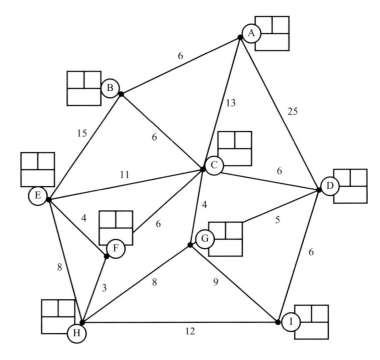

(ii) As part of a traffic management scheme it is proposed to turn the road connecting C and G into a one-way road, traffic only being allowed to proceed in the direction from G to C. What differences would this make to your shortest routes and distances?

(iii) In protest at the proposals in part (ii), and before they are implemented, a group of road users stages a demonstration in the centre of C which delays all traffic passing through C by 20 minutes.

  (a) Given that traffic in the area travels at 30 mph, explain how to adapt the network to model this information so that an application of Dijkstra's algorithm will produce the fastest journey *time* from A to F.
  (b) Find the fastest route from A to F during the demonstration.

[MEI]

9 A gardener wishes to use hoses to connect taps to a mains water supply. Taps can be connected 'in-line' ▭━●━▭ or at the end of a hose ▭━● .

Hose 'splitters' are available to enable more than one hose to leave a tap.

The nodes of the network on the next page represent the mains supply and the required positions for taps. The arcs represent possible routes for hoses, with the weights being the lengths of those routes in metres.

(i) Use Kruskal's algorithm to find the minimum length of hose needed by the gardener.

Give the total length of hose required.

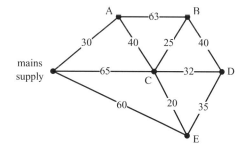

(ii) The gardener is concerned that the solution corresponding to the minimum length of hose will require too much water to pass through the hose supplying tap C. He decides to investigate other solutions involving shortest routes from the mains supply.

Use Dijkstra's algorithm to find the shortest route from the mains supply to tap D.

(iii) (a) By laying hose along the shortest route from the mains supply to tap D, the gardener finds that he does not need some of the lengths of hose indicated by the solution to part (i). This achieves a solution in which no length of hose supplies more than three taps, and in which the shortest possible total length of hose is used.

Show this solution.

(b) What extra length of hose will he need to implement the solution to part (iii)(a), compared to the solution to part (i), and what percentage increase does this represent?

[MEI]

| KEY POINTS |

1 In a network with weights which are all positive the solution to the minimum connector problem is a minimal spanning tree.

2 Kruskal's algorithm finds a minimal spanning tree by choosing edges in order of weight, least weight first, until all vertices are connected. An edge must not be chosen if it creates a cycle.

3 Prim's algorithm finds a minimal spanning tree by repeatedly adding to a connected set that (or a) vertex which is 'closest' to the connected set. (Note: it does *not* aim to connect in that vertex which is 'closest' to the last vertex which was connected.) Prim's algorithm can be implemented in graphical form and in tabular form.

4 Dijkstra's algorithm is a labelling algorithm for finding a least weight path between two vertices in a network. At each iteration the (or a) vertex with least temporary label (working value) has that label made permanent. Temporary labels on all vertices connected directly to the newly-labelled vertex are then updated if (new permanent label) + (edge weight) is an improvement. This is repeated until the destination vertex is permanently labelled. The label is the required least weight and the route is found by tracing back.

5 Kruskal's and Prim's algorithms have cubic complexity.

6 Dijkstra's algorithm has quadratic complexity. If Dijkstra's algorithm is used repeatedly to find all shortest routes in a network, then that has cubic complexity.

Critical path analysis

'Let all things be done decently and in order.'

The First Epistle of Paul to the Corinthians, xiv 40

Every morning you toast three pieces of bread under a grill. The grill will take two pieces of bread at a time and takes 30 seconds to toast each side of the bread. The schedule for the operation could be as follows.

Toast one side of pieces A and B	30s
Toast other side of pieces A and B	30s
Toast one side of piece C	30s
Toast other side of piece C	30s
Total time	120s

The schedule can be shortened, as shown in a famous wartime advertisement to encourage fuel saving, as follows.

Toast one side of pieces A and B	30s
Toast one side of C and other side of A	30s
Toast other side of B and C	30s
Total time	90s

This is a rather trivial example of how time can be saved by careful planning, but many similar opportunities arise in large-scale construction and maintenance programmes. *Critical path analysis* is a technique that enables us to plan and monitor complex projects, so that they are approached and carried out as efficiently as possible.

First, critical path analysis is looked at in the context of a simple example.

EXAMPLE 4.1

Jane, Sue and Meena share a flat.

'Why didn't you wake me?' says Jane, emerging bleary-eyed from her bedroom. 'I've got an interview at 9 o'clock and it's ten past eight already!'

'Don't worry', says Sue, 'I'll go and get your car from the garage and Meena will make you some breakfast. You go and shower and get dressed; you've got bags of time.'

Has Jane 'bags of time' with all this help or not?

Drawing the activity network

First you need to list the activities and assess how long each will take.

Shower	3 minutes
Dress	8 minutes
Fetch car (from lock-up garage nearby)	7 minutes
Make breakfast	12 minutes
Eat breakfast	10 minutes
Drive to interview	20 minutes

Next you must determine the logical relationship between the activities. Some activities can take place concurrently; for example, showering, making breakfast and fetching the car can all start immediately. Some activities cannot begin until others are completed; for example, Jane cannot dress until she has showered and she cannot eat breakfast until both she has dressed and Meena has prepared her breakfast.

You could produce a list of all the activities and their immediately preceding activities.

	Activity	Immediately preceding activities
A	Shower	–
B	Dress	A
C	Fetch car	–
D	Make breakfast	–
E	Eat breakfast	B, D
F	Drive to interview	C, E

The start/finish of one or more activities is referred to as an *event*. In this example the first event marks the start of the activities 'make breakfast', 'shower' and 'fetch car'. Another event marks the completion of 'make breakfast' and 'dress' and the start of 'eat breakfast'.

You can now draw an activity network to represent the sequence of activities.

SOME RULES FOR ACTIVITY NETWORKS (PRECEDENCE DIAGRAMS)

1 Activities are represented by arcs.

2 Events are represented by nodes.

3 Events are numbered in such a way as to enable each activity to be referred to by a unique ordered pair of event numbers i, j with $j > i$. Dummy activities are introduced if necessary to ensure unique numbering or to model the precedences. They are denoted by broken lines and have 0 duration (as, for example, in figure 4.1).

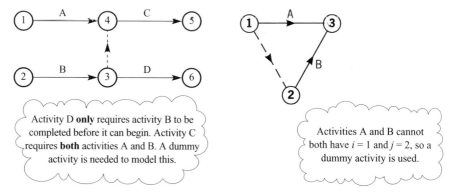

Figure 4.1

4 There should be one start and one finish node.

5 Details of the activity are written along the arc with the duration shown.

6 The length of an arc is not significant.

The activity network for Example 4.1 is shown in figure 4.2.

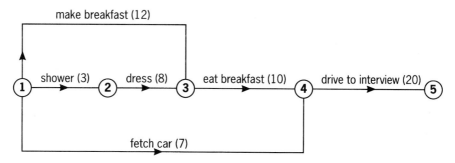

Figure 4.2

Identifying a critical path

It is clear that the duration of the whole operation is that of the longest path through the network. Any delay in the activities on this path will increase the duration of the whole operation. Such a path is called a *critical path* and the activities on it are referred to as *critical activities*.

To determine a critical path, you first move through the network from start to finish calculating the *earliest event times*. These are the earliest times that you can leave each of the events, bearing in mind that all activities leading into an event must be completed before you can leave it.

Earliest event times

The *earliest event time* for the start event is set to 0. In the example, this will correspond to 8.10 am, and all times will be given in minutes after this start time. Beside each event you draw a double box: the left-hand side will hold the earliest event time and the right-hand side the latest event time which you will meet later.

The earliest time 'dress' can begin is after 3 minutes, since Jane has to shower first and this takes 3 minutes. Although Jane can be showered and dressed in a total of 11 minutes, breakfast will not be ready until time 12, so the earliest time 'eat breakfast' can begin is 12. Whenever the start of an activity depends on two or more other activities being completed you must take the largest time so that all the relevant activities have been completed.

Continuing in this way you obtain the earliest event times shown in figure 4.3.

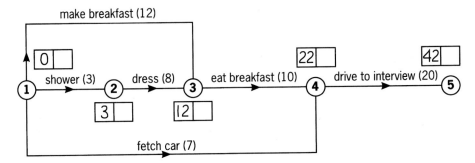

Figure 4.3

In general, to find an earliest event time, you have to consider each of the activities leading into an event. For each one, you add its duration to the earliest event time associated with its start. The earliest event time for the event under consideration is the largest of these values. The minimum completion time for the whole operation is the earliest event time for the final event.

Latest event times

The *latest event time* is the latest time you can leave an event if the operation is to be completed within its minimum completion time. Latest event times are established by working backwards through the network. First you make the latest event time for the finish event equal to its earliest event time.

Working backwards through the example, you see that the latest event time for event 4 is then 22, as Jane cannot start driving to her interview any later without delaying the operation. Similarly, for event 3 the latest event time is 12.

The latest event time for event 2 will be 4, because you could leave dressing (which takes 8 minutes) until time 4 without delaying the start of eating breakfast. There are three activities starting from event 1. You must choose the latest event time that will not lead to any delay, so you must choose the minimum of $(12 - 12)$, $(4 - 3)$ and $(22 - 7)$, which is 0.

Otherwise breakfast will not be ready on time. The resulting diagram is shown in figure. 4.4.

If you put the same numbers in the two halves of the finish event box, you should always work back to $(0,0)$ in the start event box. This is a useful check on your working.

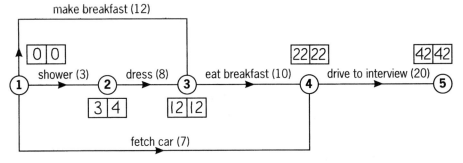

Figure 4.4

In general, to find a latest event time, you have to consider each of the activities that begin at an event. For each one, you subtract its duration from the latest event time associated with the end of the activity. The latest event time for the event under consideration is the smallest of these.

Critical activities are ones that must start and finish on time if the operation is not to be delayed. Looking at figure 4.4 it is clear that preparing and eating breakfast and driving to the interview are critical. Any delay in these activities will extend the time for the operation beyond its minimum of 42 minutes. Jane should get to her interview with 8 minutes to spare if all activities on the critical path are completed on time.

In general, if the difference between the earliest event time at the beginning of an activity and the latest event time at the end of the activity equals the duration of that activity, then the activity is critical. This fact would be useful if you were wanting to write a computer program to find critical activities.

The non-critical activities have some spare time (called *float*) associated with them. For example, fetching the car takes only 7 minutes, and the car is not required until time 22, so there can be a delay of 15 minutes without affecting the overall duration of the operation.

In general, if the difference between the earliest time at which an activity can begin and the latest time at which it can be completed is greater than its duration, the activity will have float. You will consider float in more detail next.

The completed network with critical activities highlighted is shown in figure 4.5.

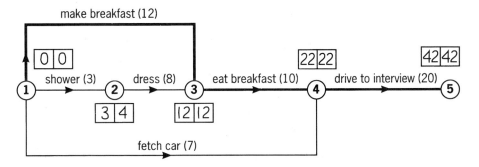

Figure 4.5

Float

You can see that there is one minute to spare for the two activities 'shower' and 'dress'. It is one minute shared between them. If Jane takes an extra minute in the shower then there is no spare time to get dressed. Similarly, if she wants to give herself that extra minute to dress, she must keep her shower to three minutes. Alternatively, she can spend an extra half minute on each activity and so on.

The spare time associated with fetching the car (15 minutes) belongs just to that activity and is referred to as *independent* float. Use could be made of this float to allow Sue to make breakfast before fetching the car. So even if Meena were not available the operation could still be completed in 42 minutes. In fact there would still be 3 minutes of spare time associated with fetching the car.

In general for activity (i, j):

$$\text{total float} = \text{latest event time for event } j - \text{earliest event time for event } i - \text{duration of activity};$$

$$\text{independent float} = \text{earliest event time for event } j - \text{latest event time for event } i - \text{duration of activity}$$
$$\text{(or 0, if the computation gives a negative answer.)}$$

$$\text{interfering float} = \text{total float} - \text{independent float}$$

 1 Can you see why these formulae give the required results?

2 The latest time that an activity can start is latest event time for event j – duration. Can you give definitions for the earliest time an activity can start and for the earliest and latest end times?

Computer programs for critical path analysis

Many computer programs are available to carry out critical path analysis. It is when using certain packages that the activity numbering convention becomes particularly important. A typical output from such a package applied to the example is given below.

		Activities		Earliest activity times		Latest activity times		Float	
i	j		Dur.	St.	End	St.	End	Total	Indep.
1	2	Shower	3	0	3	1	4	1	0
1	3	*Make breakfast*	12	0	12	0	12	0	0
1	4	Fetch car	7	0	7	15	22	15	15
2	3	Dress	8	3	11	4	12	1	0
3	4	*Eat breakfast*	10	12	22	12	22	0	0
4	5	*Drive to interview*	20	22	42	22	42	0	0

(Critical activities in italics.)

Other programs simply require a list of activities and durations together with the preceding activities. For example:

Activity		Duration	Immediately preceding activities
A	Shower	3	–
B	Dress	8	A
C	Fetch car	7	–
D	Make breakfast	12	–
E	Eat breakfast	10	B, D
F	Drive to interview	20	C, E

The output from the program is shown in figure 4.6.

```
                          Three Girls

CAN Activity              Preceding  Succeeding Activity  EST  EFT  LST  LFT
    Description           Activities Activities Duration

 1  Shower               -          2          3          0    3    1    4
 2  Dress                1          5          8          3    11   4    12
 3  Fetch Car            -          6          7          0    7    15   22
 4  Make breakfast       -          5          12         0    12   0    12
 5  Eat breakfast        2 4        6          10         12   22   12   22
 6  Drive to interview   3 5        -          20         22   42   22   42
```

Figure 4.6

1 An activity network for the cleaning of an industrial boiler is given below. The activity times are in hours. The boiler is to be brought back into service as quickly as possible. How quickly can the operation be completed and which are the critical activities?

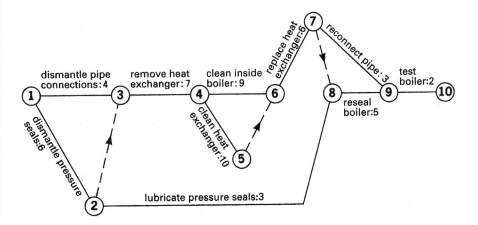

2 A construction project has been represented in network form below.

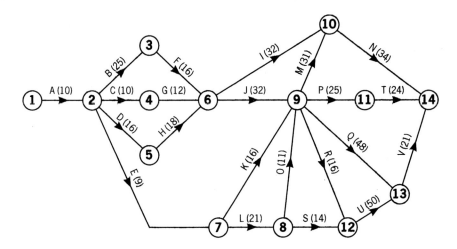

Analyse the network to determine how much time will be required to complete the project, and to identify which activities are critical.

An important use of critical path analysis is in the continuous monitoring of a project whilst it is under way. Use it to tackle the following problems.

(i) What would be the effect on the operation as a whole if a reduction in the resources available for activity H caused its duration to be increased to 28 days?

(ii) The supervisor in charge of activity I wants her team to work overtime to reduce the duration of this activity to 24 days. If you were the person in overall control of the project, how would you respond to this request?

(iii) Money is available to spend on extra machinery for either activity B or activity Q. The extra machine for activity B will reduce its duration to 20 days, a saving of 5 days. For the same outlay a machine could be purchased to cut the time of activity Q by 50%. What would you advise?

(iv) At the end of 60 days you are advised that activities A, B, C, D, E, F, G and H are completed, but that due to late delivery of materials activity K cannot be started for another 14 days. The company will have to pay a penalty if the project takes more than 175 days. Assess the effect of this delay and advise accordingly.

3 Draw a network and carry out the analysis for the following activities involved in building a house. What is the minimum completion time and which activities are critical?

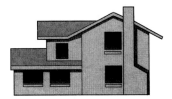

Activity		Duration (days)	Preceding activities
A	Prepare foundations	7	–
B	Order bricks	5	–
C	Order tiles	12	–
D	Lay drains	7	A
E	Erect shell	10	A, B
F	Roofing	4	C, E
G	Flooring	5	F
H	Plumbing	12	D, G
I	Glazing	1	G
J	Wiring	10	G
K	Plastering	6	H, J
L	Fittings	2	I, K
M	Clear site	2	I, K
N	Paint and clean	6	L
O	Lay paths	2	M

4 The stages involved in recording and promoting a compact disc are shown in the table below.

Activity		Duration (weeks)	Preceding activities
A	Tape the performance	10	–
B	Design the cover	9	–
C	Book adverts in press	3	–
D	Tape to CD	2	A
E	Produce cover	4	B
F	Packing	1	D, E
G	Promotion copies to radio etc.	1	D, E
H	Dispatch to shops	3	F
I	Played on radio	2	G
J	Adverts in press etc.	1	C, H, I

Draw a network and carry out the analysis to find how long the project will take and which activities are critical.

5 For an airline, an aircraft on the ground is an aircraft not earning money. However, safety is of paramount importance and this requires regular thorough maintenance. Critical path analysis has been widely used in planning such maintenance schedules.

Here is a much simpler maintenance situation for you to analyse. You take your car into the garage and fill up with petrol and have a few routine checks and jobs done. You can assume that the garage is well staffed and that you can just sit back and let their team swing into action. How long is it all going to take?

Activity	Duration (seconds)
Fill petrol tank	240
Wash windscreen	60
Check tyre pressure and inflate	180
Open bonnet	10
Check oil and top up	120
Check and fill radiator	60
Check and fill battery	60
Close bonnet	10
Pay bill	90

6 Four people are travelling by car when they have a flat tyre. Here is a list of the activities involved in changing the wheel. How long will they be delayed if they work together efficiently to change it?

Activity		Duration (minutes)
A	Locate which tyre has the puncture	1
B	Unlock boot	0.5
C	Get tool kit from the boot	1
D	Get jack from the boot	2
E	Get spare wheel from boot	2
F	Fix jack under car	2
G	Remove hubcap	0.5
H	Loosen wheel nuts	2
I	Jack up car clear of ground	3
J	Remove wheel nuts	2
K	Remove wheel from hub	0.5
L	Put spare wheel on hub	1
M	Loosely replace wheel nuts	2
N	Put punctured wheel in boot	2
O	Lower jack	2
P	Remove jack from under car	1
Q	Tighten wheel nuts	2
R	Replace jack in boot	2
S	Replace tool kit in boot	1
T	Replace hubcap	0.5
U	Lock boot	0.5

Resource allocation

EXAMPLE 4.2

When Mr and Mrs Jaffrey go out together in the car they get away more quickly than when Mr Jaffrey goes out on his own. Here are the activities involved in each case.

MR JAFFREY

Walk to garage doors	10	seconds
Open garage door	5	
Walk to car	5	
Enter and start car	10	
Drive car out of garage	5	
Walk back to garage	5	
Shut garage door	5	
Walk back to car	5	
Drive off	5	
TOTAL	55	seconds

MR AND MRS JAFFREY

MR JAFFREY		*MRS JAFFREY*	
Walk to garage doors	10	Walk to car by back door	10
Open garage door	5	Enter and start car	10
Wait for Mrs Jaffrey to drive out of garage	10	Drive car out of garage	5
Shut garage door	5	Wait for Mr Jaffrey to close garage door and get in the car	10
Get in car	5	Drive off	5
		TOTAL	40 seconds

In your critical path analysis to date you have assumed that there were enough people available to carry out concurrent activities. As you see in Example 4.2, Mr Jaffrey is kept fully occupied when he is alone, but he and his wife have some waiting time when they are both involved. Efficient planning must take account of the resources (people and equipment) required at different stages of the operation and try to make use of them in the most efficient way. You saw in Example 4.1 that Jane really only needed one helper.

The drawing of a *cascade chart*, which shows the information in the completed activity network in the form of bars drawn against a time scale, gives a different perspective on the operation and is a necessary first step in considering resource allocation. Two possible cascade charts for Example 4.1 are shown in figure 4.7. Activities are represented by bars of length proportional to their duration and critical activities are shown shaded. It has been assumed that activities start at their earliest possible times. The relationship between activities is shown by dotted vertical lines. If an activity is delayed it cannot move past one of these dotted lines: the effect will actually be to push the line to the right, along with those activities beyond it.

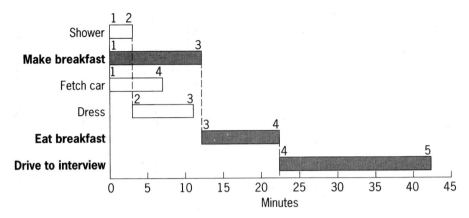

Figure 4.7(a)

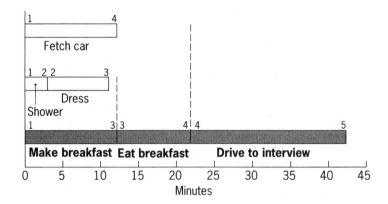

Figure 4.7(b)

In the first chart, the order of activities is that given in the computer output table on page 98, i.e. activity order (i, j) sorting first on i and then on j to give (1,2), (1,3), (1,4), (2,3), (3,4), (4,5). In the second chart, the critical activities have been put together and the others grouped appropriately.

If 'Shower' is delayed it will push 'Dress' on. This is obvious in (b) and is shown by the vertical dotted line in (a). A delay in 'Dress' longer than one minute will push 'Eat breakfast' on. The effect of a delay in 'Fetch car' is probably more clearly seen in (b): when the bar reaches the dotted line it will affect 'Drive to interview'. In (a) the dotted vertical line between the 3s appears to bar the path of this activity but in fact this link 'flies over' activity (1,4).

You can see from this that the order in which the activities are placed on the cascade chart greatly affects the ease of interpretation. For this reason, *cascade activity numbers* (CANs) are sometimes allocated to define a suitable ordering for the chart. It is easy to produce a logical order for a small number of activities by hand, but some operations have hundreds of activities and require a computer to carry out the analysis. It is these cases, as we have seen many times before, that demand an efficient algorithm such as the one shown in figure 4.8.

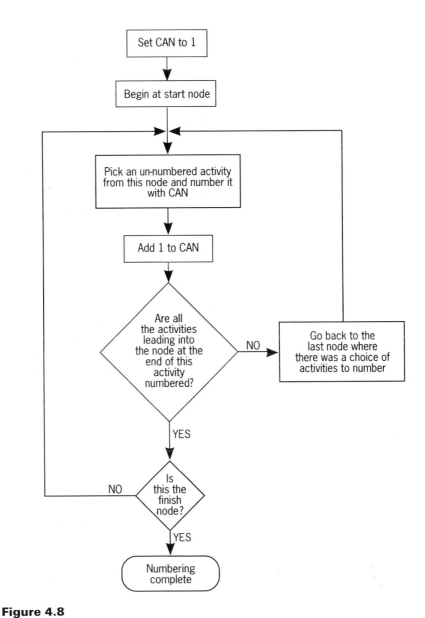

Figure 4.8

For Example 4.1 this would lead to the cascade chart shown in figure 4.9.

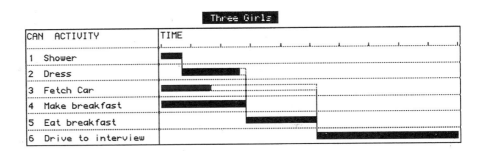

Figure 4.9

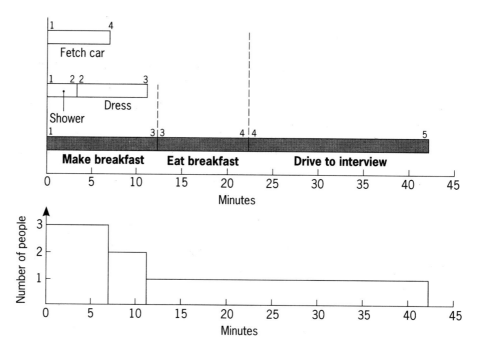

From the cascade chart it is a simple matter to draw a resource histogram to show the number of people required at any given time. Figure 4.10(a) shows the resources needed for the operation as originally defined, with all activities starting as soon as possible. In figure 4.10(b), float available on the activity 'Fetch car' has been used to smooth out the fluctuation in the resources histogram, a process referred to as *resource levelling*.

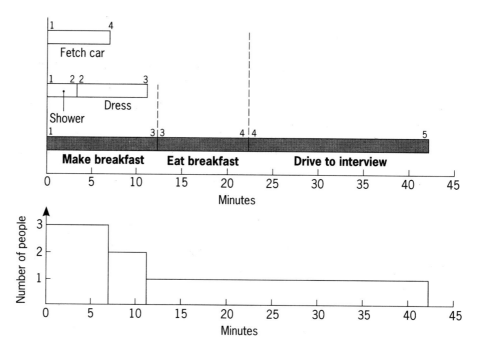

Figure 4.10(a)

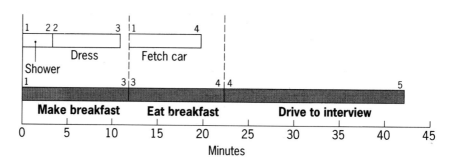

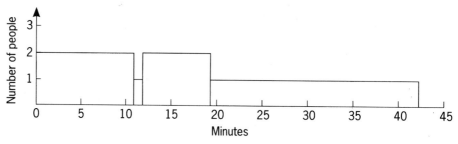

Figure 4.10(b)

The way in which resource levelling is carried out depends on what the overriding objective is. There are three possibilities, although the actual objective will often be some combination of all three.

1 MINIMISE TOTAL TIME

This is typically the case with maintenance, when you want a piece of equipment to be out of operation for as short a time as possible. For example, the overhaul of an industrial boiler or the routine maintenance of aircraft between flights would need to be planned in this way.

Using extra resources to minimise the time to completion is known as crashing a network.

2 MINIMISE TOTAL COST

When time is not at a premium, for example, when demand for a facility or piece of equipment is seasonal, it is possible to plan in this way. For example, alterations to a seaside resort hotel might be allowed to go on for a long period over the winter, so as to minimise costs.

3 MAKE MAXIMUM USE OF RESOURCES (PEOPLE AND EQUIPMENT)

If you need to work with a fixed team of workers, or if you want to avoid hiring two cement mixers when, with careful planning, one will do, then this kind of planning is appropriate.

EXERCISE 4B

1   Analyse the production of a meal of toad-in-the-hole, potatoes and cabbage, followed by apple pie and custard, the activities in which are listed below.

(i)   What is the shortest time in which it could be prepared, and how many people would be required?

(ii) If time were not critical, consider how the best use could be made of limited resources.

Activity	Duration (minutes)
Grill sausage	8
Make batter	6
Make apple pie	15
Prepare potatoes	6
Prepare cabbage	4
Cook sausage and batter together	35
Cook potatoes	25
Cook cabbage	8
Cook apple pie	30
Lay table	5
Make custard	8

2 Construct a network either by hand or using a computer package, to analyse an annual maintenance which consists of the activities listed below.

Activity	Immediately preceding activity	Duration (days)	Resources (number of workers)
A	–	1	1
B	A	2	1
C	B	4	1
D	A	3	1
E	C	14	1
F	C	14	2
G	C	16	2
H	D, G	12	2
I	D, G	14	3
J	D, G	10	1
K	H	5	1
L	I	4	2
M	J	6	1
N	E, F, K, L, M	3	2

(i) Calculate the minimum time required to complete the project and determine which activities are critical.

(ii) Calculate the amount of spare time available on the other activities, distinguishing between interfering and independent float. List those activities which are 'sub-critical', in this case defined as the ones with a float of less than seven days.

(iii) Draw a cascade chart for the operation, assuming that activities begin as soon as possible. Determine how many workers are required on each day of the project and show this manpower schedule graphically.

(iv) Describe in detail the effect on the operation if only five workers are available. You may assume that each worker can carry out any of the activities.

3 Draw and analyse the network for the project whose activities are listed below.

Activity	Immediately preceding activity	Duration (days)
A	–	8
B	–	4
C	A	2
D	A	10
E	B	5
F	C, E	3

For this project the duration of any of the activities can be accelerated, at a cost. The table below shows the normal duration and cost for each activity, the minimum time to which its duration can be reduced and the cost for this time. If an activity is accelerated to a duration greater than the minimum, the cost is calculated on a pro-rata basis, for example the accelerated cost of activity A to 7 days would be £150.

Activity	Duration	Normal cost (£)	Duration	Accelerated cost (£)
A	8	100	6	200
B	4	150	2	350
C	2	50	1	90
D	10	100	5	400
E	5	100	1	200
F	3	80	1	100

Assuming that the only possible way of reducing the total time is by increasing the costs, which activity or activities would you recommend should be accelerated if

(i) 2 days

(ii) 7 days

reduction in total time were necessary?

4 Write an algorithm to produce the (i, j) numbering of activities, given a list of the activities and those that immediately precede each one.

INVESTIGATIONS

1 Draw cascade charts and investigate the resources for questions 5 and 6 in Exercise 4A.

2 Carry out a critical path analysis for a project of your choice. Here are some suggestions:

(i) putting on a school play

(ii) giving a dinner party

(iii) starting a mini-enterprise.

ACTIVITY

In drawing cascade charts you have so far positioned the non-critical activities so that they start at the earliest possible time. If the start of an activity were delayed for a time within the available float, you would need to redraw the bar for that activity to show the new situation. If the delay were greater than the available float there would be a knock-on effect on other activities and you would need to redraw the whole chart.

To overcome this problem construct a 'dynamic' cascade chart using strips of card sliding in channels to represent the activities. The channels could be fixed on to a piece of pegboard and pegs used to mark the beginning and end of the channels in which the activities can slide. Try to devise suitable linkages to show inter-relationships between activities.

An example is shown in figure 4.11.

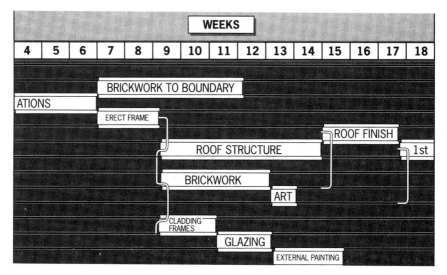

Figure 4.11

1 The table gives information about a construction project.

Activity	Duration (days)	Immediate predecessors
A	10	–
B	3	–
C	5	B
D	3	A, C
E	4	B
F	7	D
G	2	C, E

(i) Copy and complete the precedence diagram for the project.

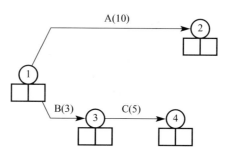

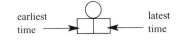
(ii) Perform a forward pass and a backward pass on your precedence network to determine the earliest and latest event times.

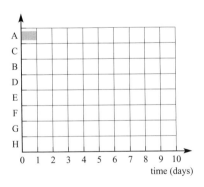

(iii) State the minimum time for completion and the critical activities.

(iv) Use an appropriate method to produce an ordering of the activities, and hence draw a cascade chart for the project.

[MEI]

2 The table shows the activities involved in a project, their durations, and their immediate predecessors.

Activity	A	B	C	D	E	F	G	H
Duration (days)	1	2	1	1	4	2	3	2
Immediate predecessors	–	–	A	B	B	C, D	E, F	E

(i) Draw a precedence diagram for the project.

(ii) Perform a forward pass and a backward pass on your precedence diagram to determine the earliest and latest event times.

State the minimum time for completion and the activities forming the critical path.

(iii) Copy and complete a cascade chart for the project, with the activities ordered as shown.

(iv) The number of people needed for each activity is as follows.

Activity	A	B	C	D	E	F	G	H
People	1	4	2	3	1	2	3	2

Activities C and F are to be scheduled to start later than their earliest start times so that only five people are needed at any one time, whilst the project is still completed in the minimum time.

Specify the scheduled start times for activities C and F.

[MEI]

3 The diagram below shows a precedence diagram. Activity durations are in days, and are shown in brackets against the arc representing the activity. Activity X connects event 6 to event 8.

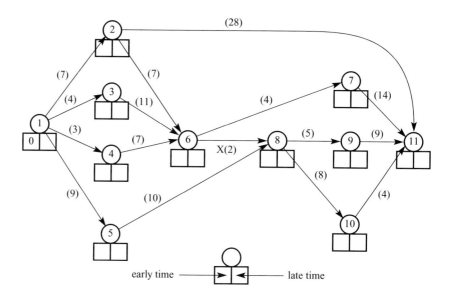

(i) Perform a forward pass on this network.

Give the early time for event number 6 and the early time for event number 8.

(ii) Perform a backward pass on the network.

Give the late time for event number 8 and the late time for event number 6.

(iii) The independent float for activity X is defined as:

early time for event 8 – late time for event 6 – duration of activity X (provided that it is non-negative).

Give the independent float for activity X.

The total float for activity X is defined as:

late time for event 8 – early time for event 6 – duration of activity X.

Give the total float for activity X.

Explain the significance of each of independent float and total float when scheduling activity X.

(iv) The network on the next page is part of a larger network. It shows all of the activities which are connected to activity X, *and only those activities*, together with some early and late times.

Say which of the following it is possible to compute from the information given on this network. In each case do the computation if it is possible, and say why it is not possible if not.

The early time for event 52.
The early time for event 58.
The late time for event 50.
The late time for event 35.

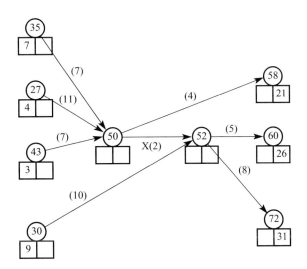

[MEI]

4 The table shows activities involved in a construction project, their durations, and their immediate predecessors.

Activity		Immediate predecessors	Duration (weeks)
A	Obtain planning permission	–	6
B	Survey site	–	2
C	Dig foundations	A, B	6
D	Lay drains	A, B	3
E	Access work	A	10
F	Plumbing	C, D	2
G	Framework	C	6
H	Internal work	E, F, G	4
I	Brickwork	G	3

(i) Draw an activity network for the project.

(ii) Perform a forward pass and a backward pass to find early and late times. Give the critical path and the minimum time to completion.

(iii) The contractor winning the contract has only one JCB (a digging machine) available. This is needed for activities C (digging foundations) and D (laying drains). Decide whether or not the contractor can complete the project within the minimum time found in (ii). Give reasons and working to support your conclusion.

[MEI]

5 (i) The table shows activities involved in a project and their immediate predecessors. It is not possible to predict exactly how long each will take, but shortest and longest possible times are shown.

Activity	Predecessors	Shortest possible time (weeks)	Longest possible time (weeks)
A	–	9	19
B	–	10	11
C	B	8	10
D	B	7	7
E	A	5	5
F	A, D	6	21
G	C, E	12	13

(a) Showing your working, find the shortest possible duration in which the entire project can be completed. List the critical activities.

(b) Showing your working, find the duration for the entire project if all activities take their longest possible times. List the critical activities.

(c) It is suggested that for each activity the shortest and longest activity times be replaced by the mean of the two, so that an estimate can be produced of the actual duration of the entire project. This leads to a project duration of 32 weeks, which is *not* equal to the mean of the shortest and longest durations in (a) and (b). Explain why not.

(ii) In the original project specification activity G could not begin until activities C and E had both been completed. A change in specification allows G to proceed when *either* C *or* E is completed.

(a) Working with the mean duration times, which are listed below, find the duration of the entire project under these new circumstances.

Activity		A	B	C	D	E	F	G
Mean of short and long times		14	10.5	9	7	5	13.5	12.5

(b) If the entire project is to be completed in the duration found in part (ii)(a), what is the earliest time at which activity C can be started? What is the latest time by which it must be finished?

[MEI]

6 The activity network shows the durations (in days) of the nine activities of a project, and their precedences.

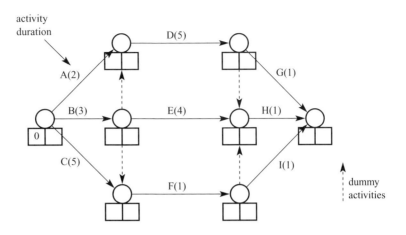

early time —— ⊞ —— late time

activity
duration

A(2)

D(5)

G(1)

B(3)   E(4)   H(1)

0

C(5)

F(1)   I(1)

dummy
activities

(i) Produce a table showing, for each activity, the *immediate* predecessors.

(ii) Perform a forward pass and a backward pass on the activity network to find the early event times and late event times.

List the critical activities and give the time for completion of the project.

(iii) Using a copy of the grid below, produce a cascade chart given that all activities are scheduled to start as early as possible. The activities have been placed in an appropriate order.

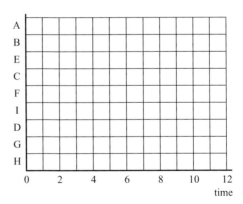

The activities require resources (people) as indicated in the table.

Activity	A	B	C	D	E	F	G	H	I
Resources required (people)	1	1	2	1	4	1	1	1	1

(iv) Produce a resource histogram using a copy of the grid provided to show resource requirements through time, given that all activities are scheduled to start as early as possible.

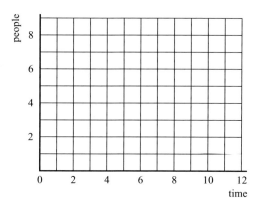

(v) If only six people are available find the shortest possible time in which the project can be completed. State which activities need to be delayed to achieve this.

[MEI]

7 (i) Tasks involved in setting an examination paper, their immediate predecessors and durations (in weeks), are given in the following precedence table.

	Task	Immediate predecessors	Duration (weeks)
A	Examiner sets questions and produces solutions	–	4
B	Printers produce first draft	A	6
C	Revisers produce recommendations	A	4
D	Examiner produces revisions to the paper	C	1
E	Examiner corrects printers' first draft, incorporating revisions	B, D	1
F	Printers produce second draft	E	4
G	Independent assessor works paper	E	5
H	Revisers examine assessor's report and printers' second draft	F, G	2
I	Examiner makes any changes that are required following assessor's report	G	1
J	Examiner considers revisers' comments and prepares final corrections for printers	H, I	1

(a) Produce an activity network for these tasks.

(b) On your diagram in (a) show the early and late times for each event. Give the critical activities and the minimum completion time for the set of ten tasks.

(ii) Versions of the examination have to be produced for two sittings a year, in January and in June, and these are produced in tandem.

-   Thus, when the examiner has completed task A for the first paper she can start task A for the second paper;
-   when the printers have completed task B for the first paper they can start task B for the second paper, etc.

(a) Complete the precedence table for the eight activities A1, B1, C1 and D1 (for the January paper), and A2, B2, C2 and D2 (for the June paper). Assume that none of the three parties involved (examiner, printers, revisers) can be working on more than one task at the same time, and that the examiner will carry out her tasks in the order A1, A2, D1, D2. (The original precedences for activities A, B, C and D still hold.)

Activity	A1	A2	B1	B2	C1	C2	D1	D2
Immediate predecessors	–					A2	C1	
						C1	A2	

(b) Complete the labelling of the activity network for the eight activities A1–D2, and show on it the early time and the late time for each event.

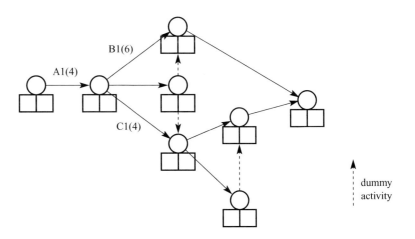

(c) How much float does the examiner have in setting the questions and producing the solutions for the second paper (activity A2)?
What is the total amount of float that she has in producing revisions for both papers (activities D1 and D2)?

[MEI]

8 On the last day of a holiday two friends want to spend as much time as possible on the beach before getting ready to catch the plane home. When they leave the beach they will need to clean up and pack, and they want to prepare and eat a meal before they go. The table gives a list of the activities which need to be completed, and their precedences.

	Activity	Immediate predecessors	Duration (minutes)
A	Walk from beach to apartment	–	10
B	Dry swimwear	A	60
C	Shower	A	5 (each)
D	Prepare meal	A	20
E	Eat meal	C, D	20
F	Clear up meal	E	20
G	Pack	B, E	15
H	Move cases to reception	F, G	5

The two showers must be taken at different times.

(i) (a) Produce an activity network representing these activities. Represent taking showers as two consecutive five-minute activities, C1 and C2.

(b) Mark the early time and the late time for each event in boxes on your activity network. Give the minimum duration and list the critical activities.

(ii) The friends can save time by not waiting for their swimwear to dry before they start packing – they can throw it into the cases at the last moment, provided that it is then dry.

(a) Show how to change the activity network to model this. (It will be sufficient for you to re-draw only that part of the network which needs to be changed.)

(b) Give the reduced duration.

(iii) Critical path analysis does not take account of the resources that are needed to complete activities. In this case the analysis incorrectly assumes that a person can shower and prepare the meal at the same time, and can pack and clear up the meal at the same time.

Furthermore the times quoted for preparing the meal (activity D) and clearing up the meal (activity F) are both in terms of 'person-minutes'. Thus, for instance, these activities would each take one person 20 minutes, or two people 10 minutes. Activities A, E, G and H require both friends to be involved for the entire duration of the activity.

Taking these factors into account complete a schedule for the activities so that they can be completed in 75 minutes.

		Friend 1		Friend 2	
	Activity	Start	End	Start	End
A	Walk from beach to apartment	0	10	0	10
B	Dry swimwear	10	70	10	70
C	Shower	10	15	15	20
D	Prepare meal				
E	Eat meal				
F	Clear up meal				
G	Pack				
H	Move cases to reception				

[MEI]

9 The table shows the durations and precedences for the six activities of a project.

Activity	Duration (days)	Immediate predecessors
A	2	–
B	3	–
C	4	A, B
D	2	B
E	4	C
F	2	C, D

(i) **(a)** Draw an activity network for the project.

(b) Mark the early time and the late time for each event in boxes on your activity network. Give the minimum duration and list the critical activities.

(ii) An extra activity, activity X, is to be incorporated in the project. The immediate predecessors for X are A and B, and it is to be completed before activity F can begin.

Give the maximum duration allowable for X without increasing the total duration of the project.

(iii) Activity X is redesigned so that it can begin as soon as activity A is completed, and so that it is no longer dependent on activity B.

(a) Produce a revised activity network for the project, incorporating the redesigned activity X.

(b) Find the maximum duration allowable for the redesigned X without increasing the total duration of the project.

(iv) Redesigning activity X so that it is completely independent of activity B is found not to be possible. However, it can be started when activity B is two-thirds complete (and when activity A is completed). Show how to incorporate this information in an activity network. You need only show that part of the network relevant to the starting of activity X.

[MEI]

10 A boat purchased in kit form is to be assembled and launched as quickly as possible. The following activities have to be performed, according to the given precedences.

Activity	Preceding activities	Duration (days)
A Sand hull	–	2
B Fit doors	–	0.5
C Cut out and fit windows	–	1
D Fit sea cocks and install toilet	A, B	1
E Fit galley	B	0.25
F Paint hull and deck	D	3
G Apply antifouling	F	1
H Fit winches	C	0.5
I Step mast	H	0.25
J Fit internal linings	C	2
K Fit upholstery	D, E, J	0.5
L Launch	G, I, K	0.5

(i) Draw an activity network to represent the problem.

(ii) Calculate the early and the late times for each event and give the earliest and latest times at which activity E may commence.

(iii) Which are the critical activities? How may they be recognised?

(iv) What is the shortest time to complete the launch of the boat?

(v) The owner does not wish to start fitting the upholstery until the deck paint is dry, i.e. until 4 days after ending activity F. Show how to modify the network to account for this.

Find the new critical path and the shortest time to complete launching.

[Oxford]

11 When Martin returns home from work, he has to prepare and eat his supper. His chicken pie and his dessert were made the night before. The pie only needs to be cooked; the dessert just needs to be removed from the refrigerator before eating.

He wishes to eat his meal in the following order:

   (a) chicken pie, mushrooms, potatoes and salad;

   (b) dessert;

   (c) coffee.

The table on the next page lists the necessary tasks. Some of Martin's precedences have been inserted (e.g. E must be finished before F is started). Other precedences will have to be allocated, bearing in mind the practicalities.

Task		Duration (mins)	Immediately preceding task
A	Set table	5	
B	Peel potatoes	5	
C	Cook potatoes	20	B
D	Cook pie	30	
E	Wash mushrooms	2	
F	Cook mushrooms	15	E
G	Prepare salad	15	
H	Dish up and eat first course	15	
I	Boil kettle	4	H
J	Eat dessert	5	H
K	Make and drink coffee	5	
L	Wash up	8	

(i) By considering the practicalities, list the immediately preceding task(s) for each task.

(ii) Draw an activity network to represent the precedences in part (i).

(iii) Using critical path analysis, calculate the shortest possible time for Martin to complete his supper, including washing up. List the critical activities.

(iv) Martin returns home at 6 pm. Produce a timetable for each task for him to prepare, eat and clear his meal.

(v) Suppose now that the pie takes only 20 minutes to cook. According to your precedence diagram, what is the new minimum time in which Martin can complete his supper? Can he actually complete the task in this time?

[Oxford]

12 The following table gives details of a set of eight tasks which have to be completed to finish a project. The 'immediate predecessors' are those tasks which must be completed before a task may be started.

Task	Duration (days)	Immediate predecessor(s)
A	3	–
B	8	–
C	7	–
D	4	A
E	2	D, B
F	6	C
G	3	E
H	2	E

(i) Produce an activity network for the project.

(ii) Find the minimum time to completion and the critical activities.

The cost associated with each task, when completed in the normal duration, is given in the following table, together with the extra cost that would be incurred in using extra resources to complete the task in one day less than the normal duration.

Task	Cost (£) Normal duration	Extra cost (£) Completion 1 day sooner
A	3000	1000
B	6000	800
C	5000	700
D	5000	1200
E	1200	600
F	5500	1000
G	2800	1100
H	1700	500

(iii) Find the minimum cost that would be entailed in completing the project in one day less time than the minimum time to completion which you found in (ii).
Give the percentage increase in cost that this involves.

[Oxford]

13 The table shows the activities involved in a project, their durations, their immediate predecessors, and the number of people needed for each activity.

Activity	A	B	C	D	E	F
Duration (days)	1	2	1	1	4	2
Immediate predecessors	–	–	A, B	B	B	C, D
People needed	1	3	2	2	2	2

(i) Draw an activity network for this project.
(ii) Perform a forward pass and a backward pass in order to find the critical path and the minimum project duration.
(iii) Schedule activities C, D and F so that the project is completed in minimum time, but so that no more than four people are needed at any one time.

[AEB]

14 A project involves six tasks as follows.

Task	Duration (mins)	Immediate predecessors
A	15	–
B	20	–
C	10	A, B
D	6	A
E	25	B, D
F	23	C

(i) Draw an activity network for the project.

(ii) By labelling your network with earliest and latest start times, find the minimum duration of the project, the critical path, and the float time for each non-critical task.

An extra condition is now imposed – task A may not begin until task B has been underway for at least 6 minutes.

(iii) Draw a new network to take account of this condition, and say what difference it makes to the completion time and to the critical path.

[AEB]

1 The stages in a critical path analysis are as follows.

(i) Prepare the network
- Make a list of the activities involved in the operation with their durations.
- Decide on the logical sequence of activities: which activities can go on at the same time, which cannot begin until others are completed.

(ii) Draw the network
- Activities are represented by arcs.
- Events, i.e. the starts and finishes of activities, are represented by nodes.
- There should be one start and one finish node for the whole project.
- Use dummy activities to model the precedences correctly.

(iii) Analyse the network
- Calculate the earliest event times by working through the network from start to finish.
- Calculate the latest event times by working backwards through the network.
- Identify the critical activities.
- Calculate the float on non-critical activities.

2
- Resource usage is examined by constructing a cascade chart and an associated resource histogram.
- Ordering the activities on a cascade chart according to their cascade activity numbers helps to see the precedences more clearly, though it will not usually be possible to achieve complete clarity.
- Given the cascade chart and the associated resource histogram, use can be made of float to schedule activities to produce a more even use of resources. This is called resource levelling.

Linear programming

Tout est pour le mieux dans le meilleur des mondes possibles.

Voltaire, Candide.

Problems in decision mathematics are often concerned with finding the best solution to a problem given constraints that must be satisfied. For instance, in Chapter 3 (Exercise 3D, question 4) you encountered the problem of finding the shortest path between Kidderminster and Cheltenham in the West Midlands.

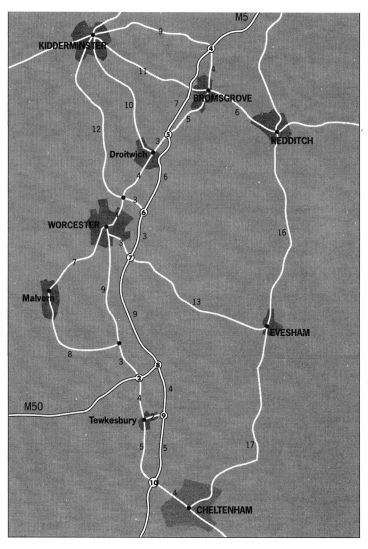

Figure 5.1 *Map to show the M5 between junctions 4 and 10*

Of course, the shortest path between two points is the straight line joining them, but without earth-boring equipment this is not a practical solution to the Kidderminster–Cheltenham problem. In the context of moving on the surface of the earth the shortest path between two points is an arc of a circle with centre at the centre of the earth and passing through the two points – a great circle. But in real life you are *constrained* to travel along roads. This is an example of a constrained optimisation problem.

In Chapter 3 you studied a specific algorithm to solve this shortest path problem. But it is also possible to use the more general technique of *linear programming*.

Historical note

Linear programming was developed in the Second World War to solve logistical problems. The 'programming' refers to a programme of activities, rather than to a computer program, although real-life linear programs (LPs) are almost always solved on a computer in practice.

The 'linear' refers to the functions which define the mathematical problems. A linear function is one in which the output is proportional to the input, for example, one in which double the input produces double the output.

Energy drink   Refresher drink

Figure 5.2

A factory produces two types of drink, an 'energy' drink and a 'refresher' drink. The day's output is to be planned. Each drink requires syrup, vitamin supplement and concentrated flavouring as shown in the table. The last row in the table shows how much of each ingredient is available for the day's production.

	Syrup	Vitamin supplement	Concentrated flavouring
5 litres of energy drink	1.25 litres	2 units	30 cc
5 litres of refresher drink	1.25 litres	1 unit	20 cc
Availabilities	250 litres	300 units	4.8 litres

How can the factory manager decide how much of each drink to make?

To answer the manager's production planning problem you will also need to know how much each drink sells for. The energy drink sells at £1 per litre. The refresher drink sells at 80p per litre.

Formulation

Translating a practical problem such as that above into the mathematical notation of a linear programming problem is known as *formulating* the problem. It is a specific example of part of a more general process known as mathematical modelling, referred to in the preface, and is the basis for your coursework.

The most important step in the formulation of an LP problem is to identify the variables which are to be used (which are known as the *decision variables*), and the most important word is **Let**.

In this case the variables are the amounts of each drink which are to be produced so:

> Let x be the number of litres of the energy drink which are to be produced.
> Let y be the number of litres of the refresher drink which are to be produced.

Note that these 'let' statements take account of the units of measure, so that the variables, x and y, are just *numbers*. In what follows, the algebraic statements are statements about *numbers* – the units are taken care of in formulating the statements, by using the definitions of x and y.

The constraints

The syrup constraint: $\quad 0.25x + 0.25y \leqslant 250$

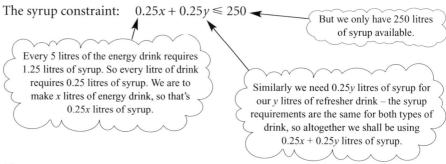

But we only have 250 litres of syrup available.

Every 5 litres of the energy drink requires 1.25 litres of syrup. So every litre of drink requires 0.25 litres of syrup. We are to make x litres of energy drink, so that's $0.25x$ litres of syrup.

Similarly we need $0.25y$ litres of syrup for our y litres of refresher drink – the syrup requirements are the same for both types of drink, so altogether we shall be using $0.25x + 0.25y$ litres of syrup.

Note

Note that this constrains your choice of *numbers* x and y. The units are taken care of in formulating the constraint. Thus, for instance, $x = 400$ and $y = 400$ would be acceptable since $(0.25 \times 400) + (0.25 \times 400) = 200$, and $200 < 250$.

But $x = 400$ and $y = 800$ would not be acceptable since $(0.25 \times 400) + (0.25 \times 800) = 300$, and $300 > 250$.

If you were submitting the problem to an LP computer package then there would be advantages in submitting the constraint as it stands, i.e. $0.25x + 0.25y \leqslant 250$. However, it will help for now to simplify it to $\boxed{x + y \leqslant 1000}$.

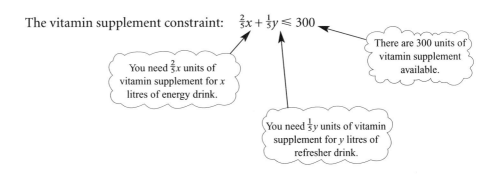

The vitamin supplement constraint: $\frac{2}{5}x + \frac{1}{5}y \leqslant 300$

You need $\frac{2}{5}x$ units of vitamin supplement for x litres of energy drink.

You need $\frac{1}{5}y$ units of vitamin supplement for y litres of refresher drink.

There are 300 units of vitamin supplement available.

 To help (for now) this simplifies to $\boxed{2x + y \leqslant 1500}$. When using a computer package there are advantages in not making this simplification.

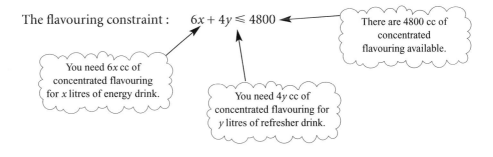

The flavouring constraint : $6x + 4y \leqslant 4800$

You need $6x$ cc of concentrated flavouring for x litres of energy drink.

You need $4y$ cc of concentrated flavouring for y litres of refresher drink.

There are 4800 cc of concentrated flavouring available.

 The simplification this time gives $3x + 2y \leqslant 2400$.

You have three constraints in total. Notice that in each case the left-hand side is a linear function of the two variables, so these are called *linear constraints*.

The objective function

In asking for the best production plan you have not yet specified what is good and what is not so good. By specifying constraints you have defined what is possible and what is not possible, but not what makes one possibility better or worse than another possibility. However, the question does specify the income from each litre of each drink, so the implication is that you are to maximise the total income.

The total income from x litres of the energy drink and y litres of the refresher drink is $1x + 0.8y$.

Letting I be income, $I = x + 0.8y$.

This is called the objective function.

Note

Here too you have a linear function of x and y – the objective is linear.

You have now completed the formulation of the practical problems as an LP expression. It would usually be written as:

$$\begin{aligned} \text{Maximise} \quad & I = x + 0.8y \\ \text{subject to} \quad & x + y \leqslant 1000 \\ & 2x + y \leqslant 1500 \\ & 3x + 2y \leqslant 2400. \end{aligned}$$

There are also two other 'assumed' constraints, $x \geqslant 0$ and $y \geqslant 0$.

Solution

When an LP has just two variables it is possible to illustrate the solution graphically. In more complex problems with many variables the logic underpinning the graphical approach can be captured algebraically, even though the corresponding many-dimensional graphs cannot be drawn.

The feasible region

The first step in constructing a graphical solution is to draw the regions represented by the inequalities.

EXAMPLE 5.1

Consider $x + y \leqslant 1000$.

SOLUTION

The set of points (x, y) whose co-ordinates satisfy the relationship $x + y = 1000$ lie on a straight line.

$x + y < 1000$ is the set of points on one side of that line, and $x + y > 1000$ is the set of points on the other side.

To illustrate $x + y \leqslant 1000$ draw the line, and then shade out the unacceptable region (figure 5.3).

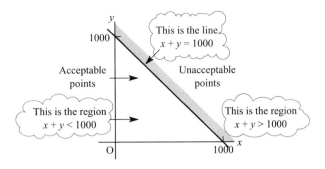

Figure 5.3

Shading *out* the unacceptable region keeps the feasible region clear and easy to identify.

 In terms of pure mathematics there is a distinction to be drawn between $x + y < 1000$ and $x + y \leqslant 1000$. A convention that is often used is that a continuous line is used for the $\leqslant$ (or $\geqslant$) case, but a broken line for the $<$ (or $>$) case. In practical problems you need not worry about whether points on the line are in or out of your acceptable region, since you can find points within the acceptable region which are as close to the line as you please. Thus LP programs will give the same results whether you use $<$ or $\leqslant$ ($>$ or $\geqslant$).

You can draw a similar graph for $2x + y \leqslant 1500$.

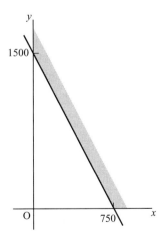

Figure 5.4

Drawing in the other constraints gives the graph in figure 5.5.

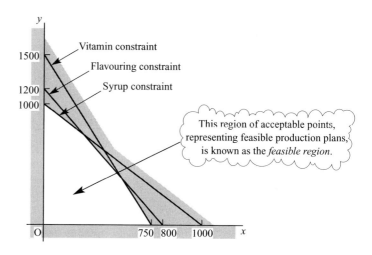

Figure 5.5

Now that you have found the set of all possible production plans you need to choose the best. In this next section you will find an argument for where to look for the best point. It is an important argument which will *motivate* methods for finding the best point: it will not be used as a technique for *finding* the best point.

On page 126 it was said that $x = 400$ and $y = 400$ satisfied the syrup constraint. Now look where the point (400, 400) is on the graph. It is in the feasible region, so $x = 400$ and $y = 400$ satisfy the other constraints as well. The income obtained from making 400 litres of the energy drink and 400 litres of the refresher drink is £400 + 0.8 × 400 = £720.

 Are there any other feasible production plans which also produce an income of £720?

There are many, all of them represented by points on the line $x + 0.8y = 720$, or rather that part of the line which lies within the feasible region (figure 5.6).

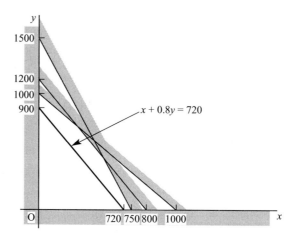

Figure 5.6

? Are there any production plans which produce an income of, say, £728. (This is equivalent to asking whether the line $x + 0.8y = 728$ crosses the feasible region.)

You can see from figure 5.7 that there are such points and that they lie on a line which is parallel to the line of points which give a profit of £720 – but further away from the origin.

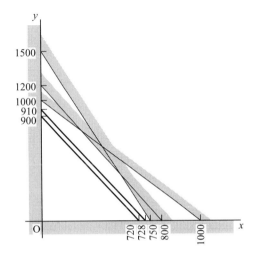

Figure 5.7

It follows that the best you can do is to find that parallel line which is as far away from the origin as possible whilst still touching the feasible region (figure 5.8).

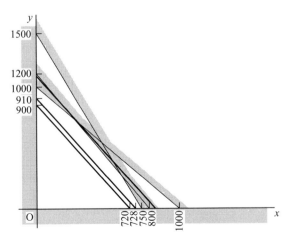

Figure 5.8

This shows that the best solution will be found at a vertex of the feasible region. By drawing a set of equal income lines you can see which vertex is best. But without actually drawing *any* lines you can see that the best result will always be at a vertex (exceptionally at adjacent vertices or points between).

Vertices are defined by the intersection of pairs of constraint lines. So it is possible to solve our LP problems by solving pairs of simultaneous equations to find the vertices, and then choosing that vertex which gives the best value of the objective function.

Note

This is how the method is translated into an algorithm for a computer program. If there are three variables then triples of simultaneous equations in three unknowns are solved to find vertices. If there are *n* variables, then *n* equations in *n* unknowns are used.

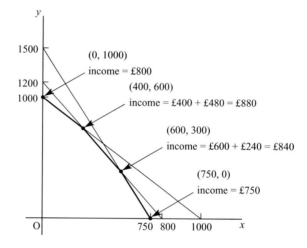

Figure 5.9

Look at the graph in figure 5.9. The income has been calculated at each vertex. The greatest value is £880 when $x = 400$ and $y = 600$.

You now have the solution of the problem. The factory should manufacture 400 litres of the energy drink and 600 litres of the refresher drink, bringing in an income of £880.

Since the best point is on the syrup constraint line and the flavouring constraint line our production plan will use all of the available syrup and all of the available flavouring. Since it is inside the vitamin constraint line there will be some vitamin supplement left.

You can check this as follows:

	Used	Left over
Syrup used = $(0.25 \times 400) + (0.25 \times 600) = 250$ litres	250	0
Flavouring used = $(6 \times 400) + (4 \times 600) = 4800$ cc	4800	0
Vitamin supplement used = $(0.4 \times 400) + (0.2 \times 600) = 280$ units	280	20

1 Solve the following LP:

maximise $P = x + 2y$

subject to $4x + 5y \leqslant 45$

$4x + 11y \leqslant 44$

$x + y \leqslant 6.$

2 A farmer grows two crops: wheat and beet. The number of hectares of wheat, x, and the number of hectares of beet, y, must satisfy

$$10x + 3y \leqslant 52$$
$$2x + 3y \leqslant 18$$
$$y \leqslant 4.$$

Determine the values of x and y for which the profit function, $P = 7x + 8y$ is a maximum. State the maximum value of P.

3 A robot can walk at 1.5 ms^{-1} or run at 4 ms^{-1}. When walking it consumes power at 1 unit per metre, and at three times this rate when running. If its batteries are charged to 9000 units, what is the greatest distance it can cover in half an hour?

4 A material manufacturer has to decide how much of each of two types of cloth to produce. Each metre of cloth A requires 2 kg of wool, $\frac{1}{2}$ litre of dye, 5 minutes of loom time, and 4 minutes of worker time. Each metre produces a profit of £3.

Cloth type B requires 1 kg of wool, $\frac{1}{3}$ litre of dye, 4 minutes of loom time, and 5 minutes of worker time. Each metre produces a profit of £2.50.

The manufacturer has available 6 hours of loom time and 6 hours of worker time for the rest of the day. There are 100 kg of wool and 28 litres of dye.

(i) Show that the constraint on loom time is redundant.
(ii) Find out how much of each cloth type the manufacturer should produce to maximise her profit.

5 A paper manufacturer has a roll of paper to cut up. It is 40 cm wide and 200 m long, and is to be cut along its length to produce widths of 11 cm for toilet rolls, and widths of 24 cm for kitchen rolls.

There are two possible cutting plans, both of which may be used on the same roll:

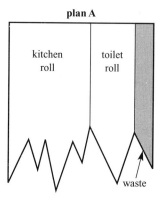

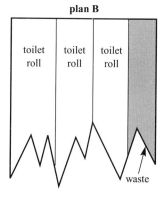

Kitchen roll paper sells for 7p a metre. Toilet roll paper sells for 4p a metre. Any of the roll that remains uncut can be sold for 8p a metre.

The manufacturer is keen to ensure that no more than 15% of the roll is wasted.

(i) Let a metres be the length that is cut to plan A, and b metres be the length cut to plan B, so that $200 - a - b$ metres remains uncut.
Formulate a linear program to find the length of roll which should be cut to each plan so as to produce the maximum income within the given constraints.

(ii) Solve the problem.

[Oxford]

6 James Bond is very particular about his cocktails. He has them mixed from gin and martini and he insists that they satisfy the following constraints:

Dryness

Gin has a dryness rating of 1 and martini a dryness rating of 3. Dryness blends linearly, i.e. a mixture of x ml of gin and y ml of martini has a dryness of

$$\frac{x + 3y}{x + y}.$$

James insists that the dryness of his cocktail is less than or equal to 2.

Alcohol

Gin is 45% alcohol by volume. Martini is 15%. Alcohol also blends linearly. James insists that his cocktail is between 18% and 36% alcohol.

James has ordered a 200 ml cocktail. Let the amount of gin in it be x ml, and let the amount of martini be y ml, so that $x + y = 200$.

(i) Explain why the dryness constraint may be expressed as $x + 3y \leqslant 400$.

(ii) In a similar way produce and simplify two further constraints.

(iii) Graph all three constraints.

(iv) Given that gin is more expensive than martini, and remembering that $x + y$ must equal 200, give the cheapest and most expensive cocktails that will satisfy James's requirements.

(v) An alternative approach to this problem is to let p be the *proportion* of gin in the cocktail, so that $1 - p$ is the proportion of martini. Using this approach the first constraint becomes $p + 3(1 - p) \leqslant 2$, i.e. $p \geqslant \frac{1}{2}$.

Express the other constraints in this way and compare the acceptable values for x with those which you obtained for x in (iv).

[Oxford]

INVESTIGATION

Find all of the solutions to the following linear programming problem:

maximise $\quad 7\frac{1}{2}x + 6y$

subject to $\quad 2x + 3y \leqslant 60$

$\qquad\qquad 10x + 3y \leqslant 150$

$\qquad\qquad 5x + 4y \leqslant 100.$

Explain why there are multiple solutions.

Does this invalidate the assertion that the best solution is always at a vertex?

(Answer provided.)

Integer programming

There are many practical situations in which the decision variables are constrained to take integer values only, in addition to satisfying linear constraints. For instance, the variables might represent the numbers of each of two types of item which are to be produced.

Theoretically and computationally this situation, integer programming, is much more difficult to deal with than is linear programming, and we shall not be dealing with algorithms which guarantee to find the optimal solution. It is, however, usually the case that the IP solution is near to the LP solution. Therefore a heuristic approach is to ignore the integer constraint, solve the associated LP, then examine integer solutions in the vicinity of the LP solution.

Note

Of course the LP solution might just happen to be integer, in which case it will be the optimal solution to the IP problem.

EXAMPLE 5.2

A landscaping project involves an area of 2000 m². Trees cost £30 each and require 30 m² of space. Shrubs cost £9 each and need 4 m² of space. At least 75 shrubs must be planted, and £3700 is available to be spent. Trees are thought to be five times as beneficial to wildlife as are shrubs. How many trees and shrubs should be planted to maximise the environmental benefit?

SOLUTION

Let x be the number of trees planted.
Let y be the number of shrubs planted.

Maximise	$5x + y$	(environmental benefit)
subject to	$30x + 9y \leqslant 3700$	(cost constraint)
	$30x + 4y \leqslant 2000$	(space constraint)
	$y \geqslant 75$	(shrubs constraint).

$$30x + 9y = 3700$$
$$\underline{30x + 4y = 2000}$$

$$5y = 1700$$
$$\underline{y = 340}$$

$$30x + 1360 = 2000$$
$$30x = 640$$
$$x = 21\tfrac{1}{3}$$

$$30x + (4 \times 75) = 2000$$
$$30x = 1700$$
$$x = 56\tfrac{2}{3}$$

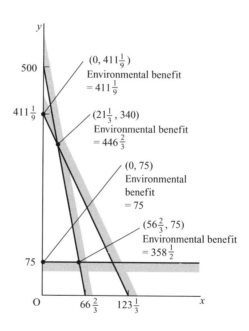

Figure 5.10

So the LP solution is $x = 21\tfrac{1}{3}$, $y = 340$ with an environmental benefit measure of $446\tfrac{2}{3}$. The obvious integer point to check is $(21, 340)$, since this is clearly feasible, and is near to $(21\tfrac{1}{3}, 340)$. It gives an environmental benefit of 445. But you can do (a little) better!

The point $(21, 341)$ is in the feasible region since

$$(30 \times 21) + (9 \times 341) = 630 + 3069 = 3699 < 3700.$$

This gives an environmental benefit of 446, and this is the (or at least an) optimal solution for this problem since 446 is the integer part of the LP solution. Had this not been the case then the optimal point might possibly have been considerably further away from the optimal LP point.

So the solution to the problem is to plant 21 trees and 341 shrubs.

1 A builder can build either luxury houses or standard houses on a plot of land. Planning regulations prevent the builder from building more than 30 houses altogether, and he wants to build at least 5 luxury houses and at least 10 standard houses. Each luxury house requires 300 m^2 of land, and each standard house requires 150 m^2 of land. The total area of the plot is 6500 m^2.

Given that the profit on a luxury house is £14 000 and that the profit on a standard house is £9000, find how many of each type of house he should build to maximise his profit.

2 A company manufactures two types of container, each requiring the same amount of material. The first type of container requires 4 seconds on a cutting machine and 3 seconds on a sewing machine.

The second type of container requires 2 seconds on the cutting machine and 7 seconds on the sewing machine. Each machine is available for 1 hour. The first type of container gives a profit of 40p. The second type gives a profit of 30p. How many of each type should be made to maximise profit?

3 A car park with total usable area 300 m^2 is to have spaces marked out for small cars and for large cars. A small car space has area 10 m^2 and a large car space has area 12 m^2. The ratio of small cars to large cars parked at any one time is estimated to be between 2:3 and 2:1. Find the number of spaces of each type that should be provided so as to maximise the number of cars that can be parked.

4 Maximise $z = x + y$
   subject to $3x + 4y \leqslant 12$
                 $2x + y \leqslant 4$
                 x integer
                 y integer.

5 A factory produces sprockets and widgets. Sprockets need 10 minutes of labour each, 15 minutes of machine time, and £20 of materials. Widgets need 20 minutes of labour each, 10 minutes of machine time and £30 of materials. For the next day there are 18 hours of labour, 20 hours of machine time and £2000 of materials.

Given that the objective is to manufacture as many completed items as possible, find how many sprockets and how many widgets should be produced.

6 A baker has 8.5 kg of flour, 5.5 kg butter and 5 kg sugar available at the end of a working day. With these ingredients, biscuits and/or buns can be made. The recipes are as follows:

30 biscuits	200 g flour	40 buns	200 g flour
	120 g butter		200 g butter
	100 g sugar		200 g sugar

(i) Biscuits are sold for 5p each and buns for 7p each. Let x be the number of biscuits baked and y be the number of buns baked. Assuming that any number of biscuits and/or buns can be baked, show that the amount of flour used (in g) is given by $6\frac{2}{3}x + 5y$.

Hence write down and simplify an inequality constraining the values of x and y.

(ii) Produce two further inequalities relating to the availability of butter and of sugar.

(iii) Assuming that all biscuits and buns can be sold, produce a function which gives the income from the sales.

(iv) Formulate and solve the linear program:
Maximise income subject to the availability of flour, butter and sugar.
Assuming it is possible to make part-batches, state the maximum income expected, and the numbers of biscuits and buns baked.

[Oxford]

Minimisation

So far all of our problems have been maximisation problems. Minimisation problems can be solved graphically in exactly the same way. In a later module, when LP problems are solved algebraically, minimisation can be more difficult than maximisation.

EXAMPLE 5.3

An oil company has two refineries. Refinery 1 produces 100 barrels of high grade oil, 200 barrels of medium grade oil, and 300 barrels of low grade oil per day. It costs £10 000 per day to operate.

Refinery 2 produces 200 barrels of high grade oil, 100 barrels of medium grade oil and 200 barrels of low grade oil per day. It costs £9000 per day to operate.

An order is received for 2000 barrels of high grade oil, 2000 barrels of medium grade oil and 3600 barrels of low grade oil. For how many days should each refinery be operated to fill the order at least cost?

SOLUTION

Let x be the number of days for which refinery 1 is operated. Let y be the number of days for which refinery 2 is operated.

High grade oil produced $= 100x + 200y$
so you require $100x + 200y \geqslant 2000$.

Medium grade oil produced $= 200x + 100y$
so you require $200x + 100y \geqslant 2000$.

Low grade oil produced $= 300x + 200y$
so you require $300x + 200y \geqslant 3600$.

Cost $= 10\,000x + 9000y$.

So the LP is:

minimise $\quad 10\,000x + 9000y$

subject to $\quad 100x + 200y \geqslant 2000$

$\quad\quad\quad\quad\quad 200x + 100y \geqslant 2000$

$\quad\quad\quad\quad\quad 300x + 200y \geqslant 3600$.

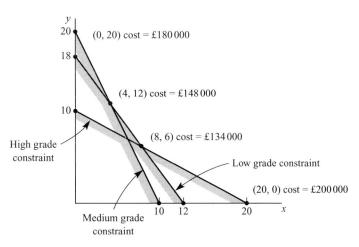

Figure 5.11

In this case, the feasible region is unbounded above, but because you have a minimisation objective, the solution will still be at one of the vertices.

Note that $(4, 12)$ is found by solving simultaneously

$\quad\quad\quad 200x + 100y = 2000$

and $\quad 300x + 200y = 3600$,

and $(8, 6)$ is found by solving simultaneously

$\quad\quad\quad 100x + 200y = 2000$

and $\quad 300x + 200y = 3600$.

The solution is to operate refinery 1 for eight days and refinery 2 for six days at a cost of £134 000.

Three or more variables and slack variables

You might be able to conceive of solving linear programming problems in three variables by graphical methods. Constraints would look like, for instance, $2x + 4y + 3z \leqslant 12$. This would have the boundary shown in figure 5.12.

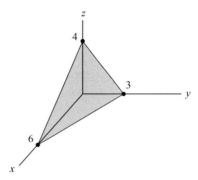

Figure 5.12

So feasible regions would be polyhedra (solids with plane faces).

Points having equal 'profit' or 'cost' would also form planes, so it can be seen that optimal points will again be at vertices. In this case, a vertex will be defined by the intersection of three planes – so you will be solving three simultaneous equations in three unknowns.

If you have more than three variables then it becomes difficult or impossible to visualise the hyperplanes which define the boundaries of the feasible region, and the hyperplanes which define the sets of points with equal profit or cost.

This can be dealt with by developing an algebraic approach which mirrors the graphical approach. The first step in this process is to establish *slack variables* to convert the constraint inequalities into equalities.

EXAMPLE 5.4

To illustrate this in the two-dimensional case consider the energy drink and the refresher drink scenario (page 128). The LP was

maximise $\qquad I = x + 0.8y$
subject to $\qquad x + y \leqslant 1000,\ 2x + y \leqslant 1500,\ 3x + 2y \leqslant 2400.$

SOLUTION

The constraints would be rewritten as

$$x + y + s_1 = 1000$$
$$2x + y + s_2 = 1500$$
$$3x + 2y + s_3 = 2400.$$

You now have three equations in five unknowns. This means in general that you can choose two unknowns to be whatever you please, and then solve for the remaining three.

Choosing two unknowns both to be equal to 0 corresponds to choosing two of the boundary lines. Calculating the values of the remaining three variables is then equivalent to finding the intersection of those two lines.

So A in figure 5.13 is given by $x = 0$ and $s_1 = 0$.

This gives
$$y = 1000 \text{ (because } s_1 = 0 \Rightarrow x + y = 1000)$$
$$s_2 = 500 \text{ (because } 2x + y + s_2 = 1500)$$
$$s_3 = 400 \text{ (because } 3x + 2y + s_3 = 2400).$$

So $A = (0, 1000, 0, 500, 400)$. ◄──── These are the values of x, y, s_1, s_2 and s_3.

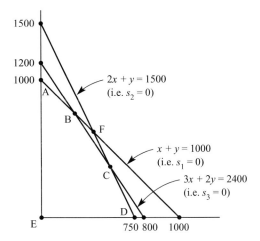

Figure 5.13

A represents a solution in which the first constraint is 'tight' (i.e. you are using all of the syrup), the second constraint has a slack of 500 (because $2x + y = 1000$ and the constraint is that $2x + y \leqslant 1500$), and the third constraint has a slack of 400.

> ⚠ Because the original constraints were simplified, the slack of 500 is not directly interpretable as units of vitamin supplement, and the 400 is not directly interpretable as litres of flavouring.

Similarly: $B = (400, 600, 0, 0, 500)$
$C = (600, 300, 100, 0, 0)$
$D = (750, 0, 250, 0, 150)$
$E = (0, 0, 1000, 1500, 2400)$.

Note

You would not wish to choose *any* two variables to be 0.

For instance, if you chose $s_1 = 0$ and $s_2 = 0$ this would define a point *F*, and would give $F = (500, 500, 0, 0, -100)$. But *F* is not feasible, and this is shown by one of the variables taking a negative value.

An algorithm known as the *simplex algorithm* provides a methodology for moving from one choice of 'zero variables' to a new and better feasible choice, i.e. one which leads to an improved value of the objective function.

Negative values

In all of our examples and exercises, variables have only been able to take non-negative values. In working with two variables, and using a graphical approach, problems in which variables are permitted to take negative values can be solved.

EXAMPLE 5.5

Maximise $\quad p = x$

subject to $\quad 3x + 2y \leqslant 6$

$\qquad\qquad x + y \geqslant 1$

$\qquad\qquad x \geqslant 0.$

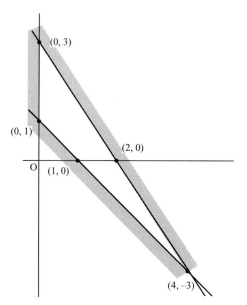

Figure 5.14

SOLUTION

$x = 4$

$y = -3.$

The simplex algorithm requires all variables to be non negative. However, if a problem requires, say, that y be allowed to take negative values, then it can be replaced in the formulation by $y_1 - y_2$ where y_1 and y_2 are both non-negative. Thus y can be negative when y_2 is greater than y_1. Software packages offer the facility for variables to be 'free' (i.e. for them to be able to take negative values), and will make the substitution automatically if the option is chosen.

The next two sections may be useful for coursework and give a taste of the scope of linear programming.

Non-linear programming

If the objective function or any of the constraints are non-linear then the problem is called a *mathematical programming problem*. The simplest case is that in which the constraints are linear, but the objective is quadratic. This special case is called a *quadratic programming problem*.

Try to solve this quadratic programming problem using a graphical approach before reading on.

Minimise $x^2 + y^2$
Subject to $x + y \geq 1$, $x \geq \frac{1}{4}$, $y \geq \frac{1}{4}$.

Note

This is a common type of problem. For instance, the manager of an investment fund may wish to hold a portfolio of shares which achieves a required return of minimum risk. This is a quadratic programming problem.

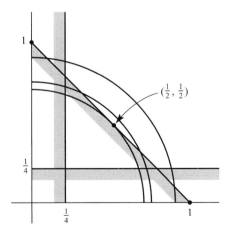

Figure 5.15

Here the feasible region is bounded (below) by linear constraints. Some objective curves of equal 'cost' are shown on figure 5.15. These curves are of the form $x^2 + y^2$ = constant. They are quarter circles, centred on the origin, which cross into the feasible region. Our task is to find the circle of smallest radius, and in particular, to find the feasible point which that smallest circle touches. Clearly, in this simple example, that is $(\frac{1}{2}, \frac{1}{2})$, which is *not* a vertex of the feasible region. In general, this makes non-linear problems much more difficult to solve.

Shortest and critical path problems

Shortest path problems can be formulated as minimisation LPs. There is a decision variable for each direction along each edge (except for returning to the start or leaving the end). The variables each take value 1 or 0 in the solution, indicating whether or not that edge is traversed in that direction. There is a constraint for each vertex indicating that if one edge is taken to arrive at the vertex then another must be taken to leave (except for the start and end vertices, for which the constraints are slightly different).

In the network shown in figure 5.16, the shortest route from A to D in the network is given by:

Minimise

$2AB + 5AC + 4BD + 2CD + BC + CB$

(This is the total distance travelled as each edge variable (AB, AC, etc.) will have value 1 or 0.)

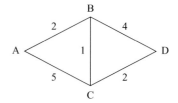

Figure 5.16

Subject to

$AB + AC = 1$ (because you must leave from A)

$BD + CD = 1$ (because you must finish at D)

$AB + CB − BD − BC = 0$ (because if you arrive at B you must leave it)

$AC + BC − CB − CD = 0$ (because if you arrive at C you must leave it)

Critical path problems are similar, but are maximisation problems and the edges are directed. For instance, the critical path in figure 4.2 on page 95 could be found as follows. Using the initial letters of the activities to denote the variables:

Maximise $12m + 3s + 8d + 7f + 10e + 20dr$

subject to $m + s + f = 1$

$s − d = 0$

$m + d − e = 0$

$e + f − dr = 0$

$dr = 1.$

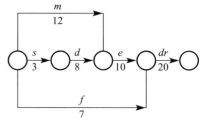

You are not expected to use LP to solve such problems at this stage since they involve more than two variables.

Figure 5.17

1 A food company is designing a new product using wheatgerm and oat flour. Nutritional requirements must be met.

Each ounce of wheatgerm contains 2 milligrams of niacin, 3 milligrams of iron and 0.5 milligrams of thiamin, and costs 8p.

Each ounce of oat flour contains 3 milligrams of niacin, 3 milligrams of iron, 0.25 milligrams of thiamin, and costs 5p.

A portion of the product must contain at least 7 milligrams of niacin, 8 milligrams of iron, and 1 milligram of thiamin.

How much of each ingredient should be used in a portion to satisfy the nutritional requirements at least cost?

2 For questions 1 and 2 of Exercise 5A

(i) re-write the constraints as equalities using slack variables
(ii) express the co-ordinates of vertices of the feasible region in terms of the values of the original variables and of the slack variables.

3 Coal arrives at a coal preparation plant from an opencast site and from a deep mine. It is to be mixed to produce a blend for an industrial customer. The customer requires 20 000 tonnes per week of the blend, and will pay £20 per tonne. Deep-mined coal has a marginal cost of £10 per tonne and coal from the opencast site has a marginal cost of £5 per tonne.

The blend must contain no more than 0.17% chlorine, since otherwise the hydrochloric acid produced by burning would corrode the boilers.

The blend must contain no more than 2% sulphur, since this burns to produce sulphur dioxide which subsequently dissolves to give acid rain. Acid rain damages the environment.

The blend must produce no more than 20% ash when burnt, otherwise the boilers will clog.

The blend must contain no more than 10% water, since otherwise the calorific value is affected.

The deep-mined coal has a chlorine content of 0.2%, a sulphur content of 3%, ash of 35%, and water 5%. The opencast coal has a chlorine content of 0.1%, sulphur of 1%, ash of 10%, and a water content of 12%. How much of each type of coal should be blended to satisfy the contract with maximum profit? Which constraint is critical, and which constraints are redundant?

4 An investor has £50 000 to invest. There are two investments which she is considering. Bond A yields 5% per annum and share B yields 7%. She does not wish to invest more than £30 000 in either option.

Formulate this as a linear programming problem and solve it graphically.

5 A furniture manufacturer produces tables and chairs. A table requires £20 worth of materials and 10 person hours of work. It sells for a profit of £15.

Each chair requires £8 of materials and 6 hours of work. The profit on a chair is £7.

Given that £480 and 300 worker hours are available for the next production batch, find how many tables and chairs should be produced to maximise the profit.

Why might the optimal solution not be a practical solution?

6 The contract conditions imposed by a coach company on hiring out a minibus are as follows:

  – the total number of passengers must not exceed 14
  – the total number of passengers must not be less than 10
  – children pay £5 each
  – adults pay £10 each
  – there must be at least as many full-fare passengers as half-fare passengers.

Find the maximum and minimum amounts the company can receive for the hire of its minibus.

7 A company needs a production plan for one of its products for the next three months. The demand for the product in each month is known and must be satisfied. Production in excess of demand in months 1 and 2 may be stored until needed in subsequent months – at a cost. Demand, production and storage costs are as follows:

	Month 1	Month 2	Month 3
Demand (units)	5	4	6
Production costs (per unit)	1	5	2

Storage costs = 3 per unit per month.

Let p_1 be the number of units produced during the first month and p_2 be the number of units produced during the second month. Then the company's requirements are modelled by the following five inequalities:

(a) $p_1 \geq 5$
(b) $p_1 + p_2 \geq 9$
(c) $p_1 + p_2 \leq 15$
(d) $p_1 \geq 0$
(e) $p_2 \geq 0$.

(i) Explain the significance of each of inequalities (a), (b) and (c).
(ii) Draw the five inequalities on a graph. What production plan is represented by each of the vertices of the feasible region?
(iii) Specify an objective function for this production problem.
(iv) Give the optimal production plan and its cost.

[AEB]

8 A clothing retailer needs to order at least 200 jackets to satisfy demand over the next sales period. He stocks two types of jacket which cost him £10 and £30 to purchase. He sells them at £20 and £50 respectively. He has £2700 to spend on jackets.

The cheaper jackets are bulky and each need 20 cm of hanging space. The expensive jackets need only 10 cm each. He has 40 m of hanging space for jackets.

(i) The retailer wishes to maximise his profit. Assuming that all jackets will be sold, formulate a linear program, the solution of which will indicate how many jackets of each type should be ordered.

You will need to give an inequality for each constraint and an appropriate objective function.

(ii) Solve the problem using a graphical method.

(iii) What would be the effect of trying to increase the order to satisfy a 10% increase in the demand for jackets? Give a reason for your answer.

[AEB]

9 A salesman wishes to carry samples of his company's products with him on visits to customers. He is limited by the weight that he can carry, and he never carries more than one of each product. However, he wishes to make best use of his visits by carrying items of the greatest total value.

(i) Suppose that he can carry 25 kg, and that there are just two products, X and Y of mass 16 kg and 10 kg respectively. Product X is worth £210 and product Y is worth £150.

Let x be the number of product X that he carries ($x = 0$ or 1) and let y be the number of product Y ($y = 0$ or 1).

(a) Express the constraint on the weight that he can carry as an inequality involving x and y. Illustrate your inequality on a graph.

(b) There are four combinations of items which the salesman can consider carrying. They are represented on the graph by the points $(0, 0)$, $(1, 0)$, $(0, 1)$ and $(1, 1)$. Which are feasible and which is optimal?

(ii) Now suppose that there is a third product, Z, with mass 13 kg and value £75.

If z ($z = 0$ or 1) is the number of product Z that he carries, then the constraint on the weight that the salesman can carry is $16x + 10y + 13z \leqslant 25$.

(a) How many possible combinations of items are there to check for feasibility?
How many of them are feasible?

(b) What is the optimal solution, and what is the total value of the items in it?

(iii) (a) In reality the company has 57 different products, with masses ranging from 1 kg to 20 kg. How many possible combinations of items need to be checked for feasibility?

(b) What is the difficulty in using the technique of linear programming to attempt to solve the salesman's problem?

[AEB]

10 A manufacturing company has a production plan in which at least 500 units are to be produced each week. The company employs two categories of employee: category A and category B. Category A employees are paid £265 per week and they each produce 13 units per week. Category B employees are paid £205 per week and each produces 10 units per week. Company policy is to have at least 45 employees producing these units.

(i) Write down a linear program to find the optimal mix of employees at the cheapest cost.

(ii) Use a graphical approach to solve your linear programming problem, ignoring for the moment the fact that the solution to the problem must be integer.

(iii) The best *integer* solution to the problem incurs a weekly pay bill of £10 235. Find this solution.

(iv) Compare and contrast the linear problem in part (ii) and the integer problem in part (iii), the solutions, and their associated costs.

[AEB]

11 Joan is producing a revision plan for the weekend for two subject areas, A and B. She has at most six hours available. She feels that she needs to work for at least two hours on each subject area. The subject areas contribute 60% and 40% respectively in the coming examination, and Joan intends to reflect that balance in the time that she allocates to revising for each area.

(i) Define appropriate variables and formulate four constraints (three *inequalities* and one *equality*) to model Joan's requirements.

(ii) Illustrate your constraints graphically. Indicate the feasible points and show that one of the constraints is redundant.

(iii) Produce alternative revision plans which satisfy the constraints and which use

   (a) as much time as possible

   (b) as little time as possible.

An alternative approach to solving this problem is to use the equality constraint to replace one variable by the other in each inequality, thus reducing the problem to one involving a single variable only.

(iv) Solve the problem by this approach, comparing the answer with that from (iii).

[AEB]

12 A company manufactures garden strimmers and shredders. Strimmers are sold to wholesalers at £25 each; shredders at £100. The company needs an income of £15 000 per week to remain solvent.

Production costs for a strimmer are £12 and for a shredder £60. The company is constrained to spend no more than £12 000 per week on production costs.

The company has fixed contracts for 200 strimmers and 50 shredders which it must deliver each week.

Let x be the number of strimmers manufactured per week. Let y be the number of shredders manufactured per week.

(i) Draw a graph to represent the four constraints under which the company must operate.

(ii) Show that the weekly profit (£) is given by $13x + 40y$.

(iii) Give the most profitable feasible production plan, and state the corresponding weekly profit.

(iv) In a particular week, cash flow problems force the company to operate at the lowest possible level of production costs. What production plan should be operated? What will be the corresponding production costs, and what profit will be made?

[AEB]

KEY POINTS

1 To formulate a problem as a linear programming problem first identify the decision variables. Do this by using statements which start, for instance 'Let *x* be the number of litres of the energy drink'. The formulation will consist of an objective and a number of constraints.

2 To solve the problem draw a graph in which each constraint is represented by a line together with shading. The unacceptable side of the line should be shaded. This leaves a feasible region. The solution of the problem will be at one of the vertices of the feasible region. These can be checked to find the best.

3 If the solution to the problem has to have integer values then points with integer values close to the optimal point can be checked. This is likely to reveal the optimal solution, but is not guaranteed to.

4 The first step in dealing with problems involving three or more variables is to introduce slack variables to convert the constraint inequalities into equalities. The values taken by the slack variables at any solution point represent the extent to which there is slack in the constraint at that point.

Simulation

But to us, probability is the very guide of life.

Bishop Joseph Butler (1756)

Suppose that a learner driver has a probability of $\frac{1}{3}$ of passing the driving test at any attempt, and that she takes the test until she passes. How may the process of taking the test be simulated by throwing a die? Would it be possible to use other equipment in the simulation instead of a die?

 Suppose there are several such drivers. Simulate the process of each driver taking the test until he or she passes. Record the number of attempts needed for each driver to pass, and display your results in a suitable form. What is the mean number of attempts needed? What is the most likely number of attempts needed?

Monte Carlo methods

The above situations could be analysed using probability theory. But many situations that you would like to analyse may well be too complicated even for sophisticated mathematical techniques. The investigation of the operation of queues is one such problem and you may need to answer questions such as the ones below.

- How should a set of traffic lights be set to give the best traffic flow?

- How should a doctor organise his surgery to minimise waiting times?

- What is the best queueing system for a bank or building society branch to operate to satisfy its customers?

It may be possible to investigate such questions by experimentation; for example, you could try various settings of the traffic lights until you get the best result. However, this may take a long time and not be the most popular way! It is more likely that a simulation would be used.

Simulation means the imitation of the operation of a system. You will probably have heard of flight simulators which are used in the training of aircraft pilots. In this chapter you will be looking at *stochastic* simulations. This means you will be studying situations where chance affects the outcome. The methods you will use are commonly called *Monte Carlo methods*. Monte Carlo is famous for its casinos and for games of chance. Monte Carlo methods are so called because some kind of random device such as a die or coin or even a roulette wheel is used to imitate chance happenings in the real world.

Random devices

In the introductory problem you used a die to simulate the taking of a driving test. You probably devised a rule, such as 'if the die shows a 5 or 6 then the learner passes the test, if the die shows a 1, 2, 3 or 4 then she fails'.

You may prefer, instead, to use another random device such as the random number generator on your scientific calculator. On most calculators this is indicated by a symbol like $\boxed{\text{Ran } \#}$, and when pressed it will produce a random decimal number between 0 and 1. A student pressed this key a few times on his calculator and obtained the following numbers:

0.611 0.539 0.068 0.468 0.084 0.304 0.175.

(These are called realisations of a uniformly distributed random variable.)

The student could therefore use the following rule to simulate the learner driver taking the driving test: 'If the random number generated is less than 0.333 then the learner driver passes the test; otherwise she fails.'

In fact this rule will give only a very good approximation to the correct probability of passing. Can you see why this is so? Can you suggest a rule to give precisely the correct probability?

Try some of the problems in Exercise 6A, either on your own or with a partner or group, before reading further. These will help you to see how to use simulations in different situations involving chance.

1 The collector's problem

A manufacturer of breakfast cereals is giving away cards with pictures of famous mathematicians on them. There are six different cards in the set, depicting the following mathematicians: Archimedes, Bolzano, Cardan, Descartes, Euler and Fermat.

There is one card placed in each packet of cereal by the manufacturer. The different cards are distributed at random, and naturally an interested student is keen to collect the set.

(i) Simulate, for example using a die, the number of packets the student will need to buy to get at least one of each card. Perform the simulation several times, recording each time the number of packets that a collector buys before having a complete set.

(ii) Display the data you obtain and find the mean number of packets bought.

(iii) How would you carry out the simulation if there were 10 cards in the set instead of six?

2 Pedestrian crossings

A pedestrian crossing in a busy city centre is studied, and it is found that, at peak periods, pedestrians wishing to cross from one side arrive at an average rate of 1 every 10 seconds. You can model this by assuming that in any 5-second period there is a 0.5 probability of 0 pedestrians arriving, and a 0.5 probability of 1 pedestrian arriving. (This assumes, unrealistically, but for the sake of simplicity, that no more than 1 pedestrian can arrive in any 5-second period.)

The first pedestrian to arrive at the crossing will press a button to request to cross. The lights then show 'Don't cross' for 25 seconds followed by 5 seconds of 'Cross'. During this 5-second period all the people waiting can cross.

(i) Use a coin to simulate the operation of the crossing for a period of about 300 seconds.
A table like the one below might help. Here, a tail (T) indicates no arrival, and a head (H) indicates one arrival in a given 5-second period. In the 'Lights' column, D = Don't cross, C = Cross.

Time	Result of random process	Size of queue	Lights
0	T	0	D
5	H	1	D
10	T	1	D
15	H	2	D
20	H	3	D
25	T	3	D
30	T	3	C
35	H	1	D

(ii) Use the results of your simulation to display data showing

(a) the number of people crossing each time
(b) the total lengths of time for which the traffic is allowed to flow freely.

(iii) There are some simple assumptions made in this model about how the pedestrians arrive and how the crossing operates once the button to cross is pushed. How might these be made more realistic?

3 If you have not yet done so, find and try out the random number generator on your scientific calculator. On most models a special key is provided, usually like this: Ran # . Pressing the key will produce a random decimal number between 0 and 1.

The Excel formula

=RAND()

will produce the same on an Excel spreadsheet.

(i) How could you use a calculator or computer to simulate throwing a coin to get heads or tails?

(ii) Try the following if you have a computer or a graphics calculator available. On an Excel spreadsheet enter:

=RAND()

in one cell, and then copy down to the same cells below.

On a graphics calculator:

| Int | (| 5 | × | Ran # |) | EXE |

and keep pressing | EXE |

Describe the output that is produced.

(iii) Modify the spreadsheet or calculator instruction to

(a) simulate the throwing of a die

(b) produce realisations of a uniformly distributed variable taking integer values between 0 and 9 (inclusive).

4 Repeat the driving test simulation from the beginning of the chapter several times, this time with a probability of success of 0.2 at each test. Work out the average number of attempts before a driver passes. Find, by experimentation, a connection between n, the average number of attempts to pass, and p, the probability of success at each attempt. If possible, compare your results with a theoretical analysis.

5 For the **Collector's problem** (question 1) investigate how C, the average number of cards collected to get a complete set, varies with N, the number of cards in the set.

6 **Doctor's surgery**

A doctor is analysing the amount of time that patients spend in her surgery waiting room. Her first appointment is at 9.00 am and appointments are made at 10-minute intervals thereafter with the last appointment at 11.20 am. She observes that the time spent seeing each patient varies from 5 to 15 minutes. Her receptionist informs her that patients tend to arrive up to 5 minutes before their appointment times and are very rarely late.

Making suitable assumptions, which should be clearly stated, simulate the operation of the surgery for three separate mornings and find the patients' average waiting time. You may like to use a table with the headings 'Appointment', 'Random number', 'Arrival time', 'Consultation starts', 'Random number', 'Consultation finishes' and 'Waiting time'.

Comment on your results.

Note

In a realistic simulation it would be necessary to gather data about patient arrivals and about consultation times. It would not be sensible to make assumptions.

7 Random walks

Random walk is a term used in physics for the movement of a particle in one, two or three dimensions, where at each stage the direction of the movement is affected by chance. Random walks are used to model the movement of molecules in gases and radioactive particles escaping through shielding. Using such methods it is possible to work out, for example, how thick the shielding on a nuclear fuel storage vessel needs to be to keep radioactivity outside it to a safe level.

In a simple random walk in one dimension we shall take two fixed points A and B 10 m apart as in the diagram below. A particle initially starts at 4 m from A and in every second it has a probability of $\frac{1}{2}$ of moving 1 m towards A and a probability of $\frac{1}{2}$ of moving 1 m towards B. The particle stops when it reaches A or B.

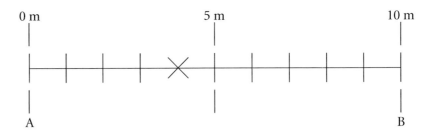

(i)   Simulate this random walk several times and hence estimate

(a) the probability of the particle reaching A
(b) the average time that the particle takes to reach A or B.

(ii) Try different starting positions.
(iii) Try different probabilities.

INVESTIGATIONS

You can tackle most of these investigations by hand. Alternatively, you may prefer to use a short program on a programmable scientific calculator or computer.

1   Use the ideas developed in Exercise 6A to simulate the operation of a real pedestrian crossing or doctor's surgery or any other similar situation. You should start by studying the situation carefully and collecting any necessary data.

A spreadsheet package could be used to help with a doctor's surgery investigation like the one in question 6 above.

2   Investigate further random walks in one dimension. Suppose, for example, that particles are emitted from the mid-point of AB. Find the average time the random walk lasts and investigate for different distances between A and B.

3   Investigate random walks in two dimensions. For example, in any second the particle might have an equal probability of moving 1 m north, south, east or west. Find how the average distance of the object from its starting point varies with the number of steps.

4   It is often said that if enough monkeys sit at enough typewriters hitting the keys at random, then they will eventually produce the complete works of Shakespeare! Investigate the validity of this statement by using a computer program to simulate one monkey sitting at one computer. For example, how long will it take, on average, for one line of Shakespeare such as '*Now is the winter of our discontent*' to be produced?

5   Suppose that in a game of tennis the server has a probability of 55% of winning any point. Simulate several games and hence estimate the probability of the server winning the game. Investigate further.

Arrival and service times

In the pedestrian crossing simulation you studied a queueing situation. You may have felt that the simple model used was not very realistic, and you probably suggested some improvements. For example, you assumed that pedestrians arrived at random at an average rate of one every 10 seconds. You modelled this by assuming that in a 5-second interval there was a 0.5 probability of one arrival and a 0.5 probability of no arrivals. Hence it was impossible to get more than one arrival in any 5-second period. This is quite a rough approximation: how could you improve it?

One obvious approach would be to consider shorter time intervals. You could keep the same average rate of arrivals by having time intervals of one second, and a probability of 0.1 for one arrival and 0.9 for no arrivals. It might be better to use still shorter time intervals such as 0.1 seconds or 0.01 seconds, with correspondingly modified probabilities: you might like to consider how short you think the intervals should be for an acceptable approximation.

If you try to re-run the pedestrian crossing simulation with time intervals of one second, you will find that to cover a 300-second time period you will need to generate five times as many random numbers as before. The simulation will therefore take about five times as long. You could possibly use a computer program to speed up the process but, like you, as the time intervals became smaller the computer would spend more actual time waiting for an arrival, so the simulation would take longer to run.

To improve both the accuracy and efficiency of a queueing simulation a different approach is needed. One possibility is to study the *inter-arrival times* or *arrival intervals*. These terms just mean the times between successive arrivals. For example, if the first pedestrian arrived at the crossing at time 2 seconds and the second pedestrian arrived at time 8 seconds then the inter-arrival time would simply be 6 seconds. You can then use random numbers to simulate directly the time between arrivals in a queue rather than look at each small time interval in turn. Try this out on the following exercise.

1 Data are collected on the inter-arrival times of cars arriving at a petrol station. The arrival interval is measured to the nearest half minute and the data is given in the table below.

Inter-arrival times (minutes)	0.5	1	1.5	2	2.5
Percentage of occasions	10%	30%	30%	20%	10%

Use random number tables or a calculator to simulate the arrival times for the first 10 cars. You might find it helpful to use a table with the headings 'Car', 'Random number', 'Arrival interval' and 'Arrival time', but leave room for about five more columns on the right.

2 Suppose that the petrol station in question 1 has just one petrol pump and that the following table gives the service times.

Service times (minutes)	1	1.5	2	2.5	3
Percentage of occasions	5%	25%	40%	20%	10%

Complete the simulation of the operation of the petrol station by adding extra columns to the table you made for question 1. Find the mean time that customers spend waiting to be served.

3 Re-run the simulation in question 2 but with two petrol pumps. Do you think two pumps are justified in this situation?

The queueing discipline

You are now almost ready to begin simulations of various queueing situations. However, a few other points are worth noting at this stage.

For any queueing system there is normally a set of rules, called the *queueing discipline*, which determines whose turn it is to be served next.

The most common queueing discipline is based on *first in, first out* or FIFO for short. 'First come, first served' is a more common expression for the same thing.

To know exactly how a queueing system operates it is also necessary to make clear the number of servers and how the customers are assigned to the servers. For the petrol pump simulation in question 2 in Exercise 6B you could say you have a single queue with one server, operating on FIFO. However, you did not make it entirely clear how the situation in question 3 might work. It may be a single queue, multiple server system as shown in figure 6.1.

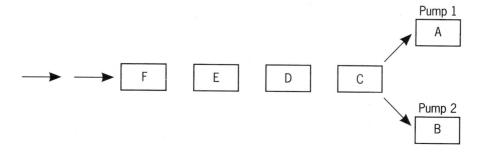

Figure 6.1

In this case the car at the head of the queue moves to the first free pump. Cars join the queue at one point at the end.

Alternatively in question 3 the two pumps might have separate queues, as in figure 6.2.

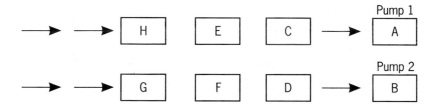

Figure 6.2

You now need to know how new arrivals decide which queue to join. Perhaps most of them would join the shortest of the two queues, or queue at random if they were of the same length.

To get reliable information from any simulation it is necessary to run it several times. Since running one simulation is a fairly lengthy and repetitive process it is normal to use a computer for simulation problems. You can do this by programming the simulation for yourself or by using a special simulation package.

You may decide to use a computer for some of the questions in the following exercise, but most of them are simple enough to be tackled by hand. Work in pairs or in a group if you wish, and attempt only a selection of the questions.

1 A petrol station has two car washes. The time taken for a car to be washed is always 12 minutes. During normal operating hours the following pattern of arrival intervals has been observed (times are measured to the nearest minute).

Interval between arrivals (minutes)	3	6	9	12	15
Frequency	30	45	15	5	5

(i) Find the average time between arrivals and hence state whether two car washes should be sufficient.

(ii) Use random number tables or a calculator to simulate the arrival times of cars for a two-hour period. Present your results in the form of a table with the headings 'Car', 'Random number', 'Arrival interval' and 'Arrival time ' but leave room for four more columns on the right for part (iii).

(iii) The car washes are served by a single queue, the car at the head of the queue moving to the first available car wash. Complete the simulation of the operation of the car washes for the two-hour period. You will need to add further columns to your table in (ii), probably with the headings 'Begins wash', 'Wash number (1 or 2)', 'Ends wash' and 'Waiting time'.

(iv) Find the mean waiting time and comment on your results.

2 A computer software company operates a telephone helpline for its customers. Calls to the helpline number are put into a queue. Recently there have been complaints of long delays before calls are answered during the busy period from 10.00 am to 12.00 noon. At the moment there is just one employee who answers the helpline calls.

As part of a study to improve customer service, data have been collected on the arrival intervals and time for each call (times are measured to the nearest minute). This is shown in the following tables.

Arrival interval (minutes)	1	2	3	4
Percentage of occasions	52	27	16	5

Call length (minutes)	1	2	3	4	5	6
Percentage of occasions	4	25	31	20	12	8

(i) Use the data to simulate the operation of the helpline for a period of about 30 minutes. Find the mean waiting time.

(ii) Do you think that more people should be employed on the helpline in the busy period? Perform an additional simulation to investigate the effect of one (or more) extra employees.

(iii) Comment on your results.

3 A small building society branch has just one serving counter. During the lunchtime hours, long queues sometimes build up. The following data were collected in order to investigate the system.

Arrival interval (minutes)	1	2	3
Percentage frequency	33	55	12

Service time (minutes)	1	2	3	4	5
Percentage frequency	8	35	34	17	6

(i) Simulate the running of the branch for a period of about 60 minutes. Find the mean waiting time for a customer.

(ii) Find from the data above the average arrival interval and service time. Do you think a second counter is justified?

(iii) Simulate the running of the branch with two counters.

(iv) Comment on your results.

4 At a fast food outlet the following pattern of arrival intervals has been observed.

Arrival interval (minutes)	0–1	1–2	2–3	3–4
Percentage frequency	30	40	20	10

(i) The fast food outlet opens at 6.00 pm each day. Use random numbers to simulate the arrival times of the first 10 customers to the nearest 0.1 minutes.

(ii) The service time can be modelled as a random time between 1 and 2 minutes. Complete the simulation of the queue for these customers.

(iii) Do you think the outlet really will provide 'fast food'?

5 The cafeteria in a local park is self-service and there is one till. Customers go to the counter, select various items, join a queue at the till and wait to receive service from the till operator. The following table gives a frequency distribution of observed inter-arrival times between 30 successive customers joining the queue at the till.

Inter-arrival time (seconds)	Number of arrivals
0–19	5
20–39	7
40–59	6
60–79	2
80–99	3
100–119	2
120–139	2
140–159	1
160–179	2
180–199	0

(i) Extend the table to include the relative frequencies and cumulative distribution of the inter-arrival times to three decimal places.

(ii) Use the following four random numbers to find a sample of four inter-arrival times from this frequency distribution:

0.176 0.241 0.427 0.144.

(iii) The following is a sequence of randomly sampled service times (in seconds):

30 20 28 32 161.

Use these, together with the inter-arrival times which you found in (ii), to carry out a simulation up to the time of the fourth arrival. Assume that the queueing discipline is FIFO and that the first person arrives at the time given by the first inter-arrival time. For how long is the queue empty from that first arrival time?

[Oxford]

6 A shop selling electrical equipment includes in its range of products a certain type of television set. The pattern of weekly demand observed for this particular television is given in the table.

Number of televisions demanded per week	0	1	2	3	4	5	6
Probability	0.15	0.25	0.25	0.15	0.10	0.05	0.05

The manager of the shop orders more televisions from the warehouse at the end of each week. These are delivered at the end of the next week, so that televisions ordered at the end of week 1 can be sold from the beginning of week 3. His current inventory policy is always to order 12 − n televisions, where n is the total number of televisions in stock at the end of the week.

(i) Simulate the operation of the inventory system for a period of 15 weeks. Set out your results in a table like the one shown.

Week	Stock at beginning	Random number	Demand	Order received	Stock at end of week
1	8			4	
2					

(ii) A simple model of the costs involved in this situation is as follows. Stockholding costs amount to £10 per week for every television in stock at the beginning of the week. Each television sold gives a net profit of £50 before stockholding costs are taken into account. Find the profit or loss made for the 15 weeks of your simulation.

(iii) Suggest an alternative inventory policy, and perform another simulation to see whether your policy would produce greater profits.
State clearly any assumptions made.

7 The manager of a petrol station is keen to improve, if possible, her inventory policy. She has supplied you with the following information.

- The station is open 6 days a week, closed Sundays.

- Demand for petrol is 1500–2500 gallons per day.

- The tanks hold 30 000 gallons in total.

- At the end of each day the amount of petrol in stock is checked. When the stock level falls below 20 000 gallons an order is placed for 15 000 gallons. The order lead-time is 4 working days, so an order placed at the end of day 1 is delivered at the end of the fifth working day.

- Stockholding costs are 0.1% per day of the value of stock at the beginning of the day. The stock value is £2 per gallon. Fixed costs are £80 per day.

- Net profit is 10p per gallon before stockholding and fixed costs are taken into account.

(i) Simulate the operation of the petrol station for a 20-day period.
Find the profit or loss made.

(ii) Suggest and investigate a different reorder policy. Petrol can be ordered in multiples of 1000 gallons, and the minimum order quantity is 10 000 gallons.

(iii) State clearly any assumptions made in your simulations and comment on your results.

Correcting rules

On page 151 you were asked why using the rule 'If the number generated on a calculator is less than 0.333 the learner driver passes...' will not accurately simulate the process in which the probability of passing is $\frac{1}{3}$.

This is because the possible outputs from the calculator will be 0.000, 0.001, 0.002, ..., 0.331, 0.332, 0.333, 0.334, ..., 0.997, 0.998, 0.999.

A careful count will reveal that there are 1000 possibilities, and that 333 of them are less than 0.333. Thus the probability achieved by the rule is $\frac{333}{1000}$, which is not quite $\frac{1}{3}$.

To correct this you could use the rule:

- If the number generated is less than 0.333 then the learner passes the test.

- If the number generated is greater than or equal to 0.333, and less than 0.999, then the learner driver fails the test.

- If the number generated is 0.999, generate a fresh number, and use that instead.

You might think that this is being unnecessarily meticulous, but the same principles apply in situations in which the effect can be *very* significant.

For instance, suppose you wish to simulate the propagation of a plant from seeds in which the probability of a seedling producing a plant with white flowers is $\frac{1}{2}$, with pink flowers is $\frac{1}{3}$ and with blue flowers is $\frac{1}{6}$.

Suppose that you have a mechanism for generating two-digit random numbers.

Then an efficient rule would be:

Random number	Result
00 – 47	White
48 – 79	Pink
80 – 95	Blue
96, 97, 98, 99	Reject and re-draw

Other rules, including less efficient rules, are possible, but all should reflect the *correct* probabilities.

INVESTIGATIONS

1 Investigate a real queueing or stock control system. You will need to collect some data on the operation of the system and you should then set up a simulation. Suggest and investigate changes that may improve the system.

2 Write a computer program to simulate a simple queueing situation.

3 Use a spreadsheet package to simulate a simple stock control problem.

EXERCISE 6D

1 Random number lists for this question are printed on the next page. The random numbers are to be used as two-digit random numbers, and are to be read in order from the top left across the page, a row at a time, as required. There are separate lists for parts (i) and (ii) of the question.

(i) A small post office has one server. Customer inter-arrival times follow the following distribution:

Inter-arrival time (minutes)	2	3	4
Probability	$\frac{1}{4}$	$\frac{1}{2}$	$\frac{1}{4}$

Using two-digit random numbers from the list marked A, simulate *five* inter-arrival times.

Describe the rule that you used to generate your inter-arrival times.

(ii) Service times have the following distribution:

Length of service (minutes)	1	2	5
Probability	$\frac{1}{3}$	$\frac{1}{2}$	$\frac{1}{6}$

Using two-digit random numbers from the list marked B, simulate service times for *six* customers.

Describe the rule that you used to generate your service times.

(iii) Assuming that the service of the first customer has just begun, simulate the service of six customers. Number the customers from 1 to 6, and record your results in a copy of the table.

Customer number	Arrival time	Start of service	End of service
1	0	0	
2			
3			
4			
5			
6			

(iv) Compute the mean queueing time (i.e. the mean of the times for which each customer queues), the mean length of queue (i.e. the total time spent queueing divided by the total elapsed time) and the server utilisation (i.e. the percentage of elapsed time for which the server is busy).

(v) Say how the reliability of the results from your simulation could be improved.

Random number list A

42	22	03	80	19	37	62	93	55	12
12	07	46	91	63	28	20	37	92	83
45	37	64	58	55	07	14	12	75	43
45	94	74	21	56	64	39	54	64	19
23	36	78	53	46	85	10	19	09	05

Random number list B

35	92	78	98	16	06	31	34	82	17
64	54	38	03	17	19	20	91	76	73
36	49	54	84	87	69	46	21	09	70
75	24	15	13	96	50	02	04	56	87
56	62	39	53	81	04	24	25	71	61

2 The weather bureau in a particular country defines each day to be either wet or dry.

Records show that if the weather today is dry then the probability that it will be dry tomorrow is $\frac{4}{5}$, and the probability that it will be wet tomorrow is $\frac{1}{5}$. If the weather today is wet then the probability that it will be wet tomorrow is $\frac{2}{7}$, and the probability that it will be dry tomorrow is $\frac{5}{7}$.

Future weather is to be simulated. Each day a two-digit random number is to be used, together with a simulation rule based on that day's weather, to simulate the weather for the next day.

When the weather is dry the rule to be used is as follows:

00–$79 \rightarrow$ weather tomorrow is dry
80–$99 \rightarrow$ weather tomorrow is wet.

(i) Give a simulation rule to simulate the weather for a day following a wet day.

(ii) The weather today is *dry*. Use the rules, together with the random numbers below, to simulate the weather for the next 14 days. Read the two-digit numbers from left to right.

Random numbers:

39	16	44	89	01	56	90	99	11	37	47	84	29	52	21
06	39	43	06	42	82	52	16	39	89	58	61	74	93	82

(iii) Use the 14 results from your simulation to calculate an estimate of the overall proportion of wet days.

(iv) Use the 14 results from your simulation to calculate an estimate of the probability of the weather tomorrow being the same as the weather today.

(v) Give two ways in which the simulation of the weather could be improved.

[MEI]

3 The time intervals between cars travelling along a one-way road are distributed as follows.

Gap (seconds)	Less than 4 seconds (mean = 2 seconds)	Between 4 seconds and 10 seconds (mean = 7 seconds)	Between 10 seconds and 60 seconds (mean = 35 seconds)
Probability	$\frac{1}{2}$	$\frac{1}{6}$	$\frac{1}{3}$

(i) A student uses a table of random numbers to simulate the time intervals between cars, using the rule:

$00, 01, 02 \rightarrow$ 2 seconds
$03 \rightarrow$ 7 seconds
$04, 05 \rightarrow$ 35 seconds
06–99 ignore.

Explain why this is an inefficient use of the random number table.

(ii) Give a rule for using two-digit random numbers, using the *mean* time intervals that are given, to simulate time intervals between cars. You should give an efficient rule.

(iii) Use your rule, together with the list of two-digit random numbers given below, to simulate the 15 time intervals between 16 cars. Read the random numbers from left to right.

Interval	1	2	3	4	5	6	7	8	9	10	11	12	13	14	15
Length (seconds)															

Random numbers:

98 01 11 49 58 52 83 43 14 82 65 21 21 89 23 74 37

Other cars arrive from a side road at a junction with the main one-way road. They have to turn left to join the flow of traffic along the main road. A car arriving at the junction can turn out of the side road and on to the main road only when there is a gap of at least 10 seconds before the next car travelling along the main road passes by.

When a car turns out of the side road it takes 4 seconds for any car queueing immediately behind to move up to the junction.

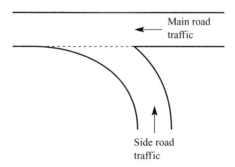

(iv) The side road serves an industrial estate. Assume that in the late afternoon cars leave the estate at a steady rate of 6 per minute (one every 10 seconds exactly).

Using as many of your simulated time intervals between cars on the main road as are needed, simulate the arrival and departure times of six cars arriving at the junction along the side road. Start your simulation as the first car arrives at the junction from the side road, and assume that at that instant a car passes the junction along the main road.

Car number	1	2	3	4	5	6
Time of arrival at junction (s)	0					
Time of departure from junction (s)						

(v) During the first 60 seconds of your simulation, what is the maximum length of the queue of cars from the side road at the junction? Give the time interval during which the queue is at its maximum length.

(vi) Suggest two ways in which the simulation might be improved.

[MEI]

4 When two friends arrive at an airport there are two check-in desks operating for their flight. There is a separate queue for each desk. There are three groups of people in the queue on the left and two groups in the queue on the right, and in each case the first group in the queue has just begun the check-in procedure.

The distribution of check-in times for groups at a check-in desk is shown in the table.

Time taken to check in (mins)	1	2	3	4	5
Probability	0.1	0.4	0.2	0.2	0.1

(i) Copy and complete the table to produce a rule for simulating group check-in times.

Random numbers	Time to check in (mins)
00–09	1
	2
	3
	4
	5

(ii) Copy and complete the following table by simulating six times what might happen if the two friends were to join the queue on the right, i.e. the shorter queue.

The friends' waiting time is the total time taken for the two groups in front to check in.

Simulation number	Random number	Time for first group to check in	Random number	Time for second group to check in	Total waiting time
1	73		91		
2	38		07		
3	03		64		
4	24		10		
5	53		70		
6	38		70		

Give the mean waiting time from your six simulations.

(iii) Instead of both joining the shorter queue the two friends decide to use the following strategy. They will join one queue each. The first to be served will then check both in.

(a) Copy and complete the table below by simulating six times the progress of the longer queue.

Simulation no.	Random no.	Time for first group to check in	Random no.	Time for second group to check in	Random no.	Time for third group to check in	Total waiting time
1	21		26		67		
2	32		73		85		
3	86		28		74		
4	60		77		95		
5	35		41		04		
6	32		93		45		

(b) Combine the results with those from the simulation in (ii) to obtain six values for the friends' waiting time if they use this strategy.

Simulation number	1	2	3	4	5	6	7	8	9	10
Waiting time (minutes)										

Give the mean of the six waiting times.

(iv) Give three ways in which this simulation experiment could be improved or made more realistic.

[MEI]

5 A 'drive-through' fast-food restaurant has a single lane for cars, alongside which are three windows. Two are for taking orders and one is for serving food. Once cars have entered the system no overtaking is possible.

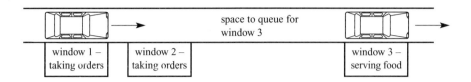

A car arrives, queues if necessary, and is then sent either to window 1 or to window 2, where the driver places an order and pays. Having paid, the driver then moves forward to window 3 as soon as possible. At window 3 the food has to be collected, which may involve a wait.

(i) The time taken to order and pay (either at window 1 or window 2) has the following probability distribution.

Time (minutes)	$\frac{1}{2}$	1	$1\frac{1}{2}$	2	$2\frac{1}{2}$
Probability	$\frac{1}{10}$	$\frac{4}{10}$	$\frac{2}{10}$	$\frac{2}{10}$	$\frac{1}{10}$

(a) Complete the table to give a rule for using two-digit random numbers to simulate times for ordering and paying.

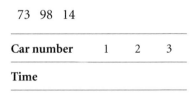

Random numbers	00 –				
Time (minutes)	$\frac{1}{2}$	1	$1\frac{1}{2}$	2	$2\frac{1}{2}$

(b) Use the following two-digit random numbers to simulate order and paying times for three cars.

73 98 14

Car number	1	2	3
Time			

(ii) The time taken from paying until the food is ready for collection at window 3 has the following probability distribution.

Time (minutes)	$\frac{1}{2}$	1	$1\frac{1}{2}$
Probability	$\frac{1}{3}$	$\frac{1}{2}$	$\frac{1}{6}$

(a) Complete the table to give a rule for using two-digit random numbers to simulate these times. (Your rule should use as many two-digit random numbers as possible.)

Random numbers	00 –		
Time (minutes)	$\frac{1}{2}$	1	$1\frac{1}{2}$

(b) Use the following two-digit random numbers to simulate three of these times: 30 24 98 49.

Order number	1	2	3
Time			

(iii) Use your times from parts (i) and (ii) to simulate the time it takes to process three cars through the system using two different rules. The arrival times of the cars have been simulated for you. Start with all windows free. (The time taken for a car to move between windows should be ignored.)

Rule 1

If both window 1 and window 2 are free the next car is sent to window 2.
If window 1 is free but window 2 is occupied the next car is sent to window 1.

Car	Arrival time (hrs:mins)	Window 1 or 2?	Times of		
			Arriving at window 1 or 2	Completing payment	Leaving window 1 or 2	Leaving system
1	12:00				
2	$12:01\frac{1}{2}$				
3	12:02				

Rule 2

If both window 1 and window 2 are free the next car is sent to window 2.
If window 1 is free and window 2 is occupied, it is sent to window 1 if the car at window 2 has been there for 1 minute or less, otherwise the next car is made to wait until window 2 is free.

Car	Arrival time (hrs:mins)	Window 1 or 2?	Times of			
			Arriving at window 1 or 2	Completing payment	Leaving window 1 or 2	Leaving system
1	12:00					
2	$12:01\frac{1}{2}$					
3	12:02					

(iv) Comment on your results from (iii).

[MEI]

KEY POINTS

1 Uniformly distributed random numbers are needed to build simulation models. Rules are constructed so that uniform random variables can be used to model observed probabilities which are not uniformly distributed.

2 To simulate possible outcomes using two-digit random numbers you need a rule in which sets of random numbers occur with the same probabilities as the outcomes which you are simulating. This may involve you rejecting some two-digit random numbers.

3 Arrival times into a queueing system are simulated by simulating the time intervals between arrivals.

4 Queueing disciplines include separate queues for separate servers and 'first in, first out' single queues.

5 Simulations need to be repeated so that the average result correctly reflects what may be expected on average. The more repetitions, the greater the confidence that may be placed in this.

6 Simulation models should be compared with real life to check their validity before using them to investigate 'what if' scenarios.

Answers

Chapter 1

❷ (Page 3)

The answer depends on the type of calculator used.

❷ (Page 7)

See text that follows.

Exercise 1A (Page 9)

1 5 slots are needed

3 1 dm

Investigations (Page 9)

The bin packing problem: using first-fit, six racks are required. The other two methods succeed with five racks. The plumbing problem: using first-fit, six lengths are required. The other two methods succeed with five lengths. The ferry loading problem: the total length of all the vehicles is 82 m, so they cannot all be loaded.

Exercise 1B (Page 11)

1

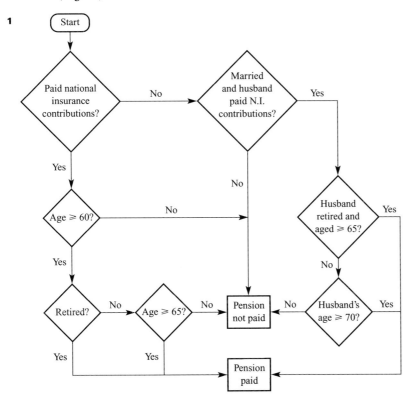

3 There are 592 bus tickets in the sequence whose digits sum to 21.
The first algorithm performs 30 592 additions (not counting the loop counters) and 10 000 comparisons. The second algorithm performs 10 592 additions, 1110 subtractions and 10 000 comparisons.

Investigations (Page 13)

3 The next two perfect numbers are 28 and 496.
4 Choose items 2, 3 and 5 with value 66.

Exercise 1C (Page 19)

1 If written in year/month/day order then the later the date the larger the number (provided that the dates do not straddle 1/1/2000).

❷ (Page 24)

With $2^n - 1$ items the list is repeatedly split into equal halves by considering the middle item.

Items:	3	7	15	31	63
Max. no. of comparisons:	1	2	3	4	5

Exercise 1D (Page 25)

3 Binary search.

4 (i) H.C.F. = 180

  (ii) One extra iteration in which the numbers are swapped.

  (iii) Algorithm requires 12 iterations. Formula gives 12.001437...

5 (i) $q = 13$, $r = 2$, 13 repetitions.

  (ii) Division (by repeated subtraction).

  (iii) $q_1 = 1$, $q_2 = 3$, $r = 2$. One execution of box 4 and three of box 7.

  (iv) Second algorithm is a form of long division. More efficient, but less transparent.

6 (i) Comparisons: 5 4 3 2 1 = 15
       Swaps: 3 3 2 1 0 = 9

  (ii) 6 comparisons and 3 swaps.

  (iii) Comparisons: 3 3 3 = 9
       Swaps: 2 2 2 = 6
     Thus total numbers of comparisons and swaps is the same under both approaches.

  (iv) Do not need to check first two pairs on first pass or first pair on second pass. By third pass list is guaranteed to be sorted.

7 (i) **(a)** 1.2 0.7
         1.1 0.4 0.4
         0.3 0.3 0.2

    (b) 1.2 0.4 0.4
         1.1 0.7 0.2
         0.3 0.3

  (ii) **(a)**

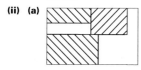

    (b)

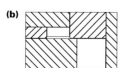

    (c) Sort by area.
       Apply algorithm in order of decreasing area.

    (d)

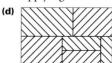

8 (i) **(a)** Comparisons – 1; 2; 2; 4; 2; 4 – 15 in total.
    (b) Comparisons – 0; 1; 1; 3; 2; 5; 4 – 16 in total.

  (ii) {1, 2, 3} 2 using shuttle and 3 using insertion;
    {1, 3, 2} 3 and 3;   {2, 1, 3} 2 and 3;
    {2, 3, 1} 3 and 2;   {3, 1, 2} 3 and 3;
    {3, 2, 1} 3 and 2.

  (iii) **(a)** e.g. {7, 6, 5, 4, 3, 2, 1} requiring 21 comparisons.

    (b) e.g. {1, 2, 3, 4, 5, 6, 7} requiring 21 comparisons.

    (c) 45 comparisons.

9 (i) **(a)** binary search.

    (b) 16. ($2^{16} < 100\,000 < 2^{17}$)

  (ii) **(a)** K4

    (b) K6

    (c) No exact halving available.

    (d) 7 guesses (8 if the answer is counted as a guess).

  (iii) 9 guesses (10 if the answer is counted as a guess).

10 (i) **(a)** 12 comparisons.

    (b) 21 comparisons.

  (ii) **(a)** ACABACECDCEGEF – 13.

    (b) ACEGEFECDCAB – 11.

    (c) Go north before east, etc. ABACDCEFEG – 9.

11 (i) **(a)** 40

    (b) 121

  (ii) 1, 1.1, 1.1.1, 1.1.2, 1.1.3, 1.2, 1.2.1, 1.2.2, 1.2.3, 1.3, 1.3.1

  (iii) 1, 1.1, 1.2, 1.3, 1.1.1, 1.1.2, 1.1.3, 1.2.1, 1.2.2, 1.2.3, 1.3.1

  (iv) earlier: 1.2, 1.3 later: 1.1.1, 1.1.2, 1.1.3, 1.2.1, 1.2.2, 1.2.3 no difference: 1.1, 1.3.1, 1.3.2, 1.3.3

  (v) Essentially different searches give different permutations of the locations, and by a symmetry argument one permutation is as good as another.

12 (i) **(a)** 48m; 56m; 43m; 25 vehicles.

    (b) Try first to place into a lower numbered lane. Places vehicle 20 into lane 2, and allows number 26 to be loaded.

    (c) In lane 1, vehicles 4, 7, 8, 14, 18 and 22, leaving 7 m.
     In lane 2, vehicles 1, 2, 3, 6, 10, 12, 15, 19, 23 and 24, leaving 3 m.
     In lane 3, vehicles 5, 9, 11, 13, 16, 17, 20, 21 and 25, leaving 3 m.

    (d) Don't stop at the first vehicles that will not fit, try to fill the lanes with others later in the queue.

  (ii) **(a)** In lane 1, vehicles 8, 18, 22, 1, 3, 9, 12, 20 and 25 leaving 0 m.
     In lane 2, vehicles 5, 6, 15, 19, 2, 4, 10, 13, 23 and 27, leaving 0 m.
     In lane 3, vehicles 14, 17, 21, 26, 7, 11, 16 and 24, leaving 1 m.

    (b) Unfairness – long vehicles arriving later may be loaded in place of shorter vehicles arriving earlier.

13 (i) Identifies square numbers.

   (ii) Squares have an odd number of (improper) factors, and thus an odd number of changes from 0 to 1. For non-squares, factors *always* come in pairs.

   (iii) e.g. AKMIMLILMKAFCDBDCBCEI* JHGHJGJFJ*IECFA a path connecting start, centre and exit.

Chapter 2

❷ **(Page 40)**

  1 A line joining two symbols means that the situation represented by one of the symbols can be transformed into the situation represented by the other symbol by one crossing of the river.

  2 Figure 2.1 shows all such linked legal situations.

❷ **(Page 43)**

This is only possible if you start at A and end at C, or vice versa.

❷ **(Page 47)**

The instructions might start: 'Send the sack of potatoes down. Next send the child down, counterbalanced by the sack of potatoes coming up. Now send the woman down, counterbalanced by the child coming back up. The potatoes are then sent to the bottom so that they can counterbalance the child going down again. The man can then descend with the woman and child coming up as counterweights…' etc.

Exercise 2A (Page 47)

1 e.g.

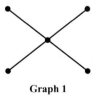
Graph 1

Graph 2

Graph 3

Graph 4

2 (i) ABCA, ABCDA, ABCDEA, ACDA, ACDFA, ADEA

  (ii) ABCEDCA is not a cycle since vertex C is repeated.

  (iii) A trail. It is a walk too, but since no edge is repeated it is best described as a trial.

3 e.g.

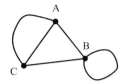

Note that the loop from B appears as a 2 in the incidence matrix as it can be followed in either direction.

4 Each edge (including a loop) contributes 2 to the vertex order total. So the vertex order total is even. So the number of vertices with odd order must be even.

5 G1: least = 3;   greatest = 6
   G2: least = $n - 1$;   greatest = $^nC_2$

6 (i) Select C and E – value 14.

  (ii) $62 - 32 = 30$ more vertices.

  (iii) Circuit not possible – all decisions are bifurcations.

    (b) Graph should be connected – all possibilities are accounted for by the branching process.

7 (i) Vertices represent areas of Kaliningrad. Edges represent bridges connecting areas.

  (ii) The pencil has to visit each vertex an even number of times.

  (iii) In an Eulerian graph, all vertices are even so, if you enter a vertex on an edge, there will be an unused edge to leave along and you can traverse the network starting and finishing at any vertex. However, if there are two odd nodes it is still possible to trace along each edge of the graph once and only once without lifting the pencil from the paper as you may leave the starting vertex one more time than you enter it and enter the finishing vertex one more time than you leave it. Note that this means that you must start at one odd vertex and finish at the other.

8 (i) $6 - 21$; $4 - 14$; $2 - 7$

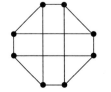

  (ii) e.g. d = 2, 4, 5, 6 or 7

9 There are five ways (including that which is given). The other four arrangements are:

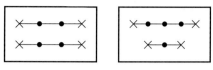

10 e.g.

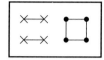 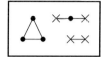

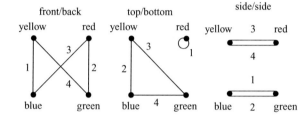

11 (i) **(a)** a–c1; b–c2; c–c3; d–c1; e–c4

   (b) a–c1; b–c2; c–c3; d–c2; e–c1

   (ii) **(a)** The vertex would have to be coloured differently to itself!

   (b) One edge is enough to force a different colour. A second adds nothing.

   (iii) **(b)** a–c1; b–c2; c–c1; d–c2; e–c3; f–c4

   (c) colour = hour; a & c then b & d then e then f

   (d) e.g. a & c then b & e then d & f

Exercise 2B (Page 52)

1 e.g. $A \leftrightarrow 1$; $D \leftrightarrow 2$; $E \leftrightarrow 3$; $B \leftrightarrow 5$; $C \leftrightarrow 6$; $F \leftrightarrow 4$

e.g.

2 (i) e.g. 1 3 2 4 5 6 1

   (ii) e.g.

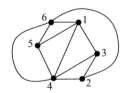

3 (i) 24

   (ii) 49!

Approx. 1 928 850 000 000 000 000 000 000 000 000 000 000 000 000 000 000 years (to 6 sf)

4 (i) AB; ACB; ADB; ACDB; ADCB

   (ii) 16

5 Draw a complete graph on 6 vertices. Imagine colouring the edge connecting two vertices green if the people represented by the vertices know each other, and red if they do not.

Choose a vertex. It has 5 edges incident upon it, so there must be three of the same colour, c_1 say. Now examine the three edges joining the vertices at the other ends of the those three. If they are all coloured c_2, then there is a c_2 triangle. If not then at least one is coloured c_1, and that means that there is a c_1 triangle which includes the original vertex.

6 Every person must map to a number. If there are n people then the image set is $\{0, 1, 2, ..., n-1\}$. But both 0 and $n-1$ cannot both be images since it is not possible simultaneously to have had somebody shake hands with everybody, and somebody else shake hands with nobody. Thus there are $n-1$ possible images for n subjects, so the mapping is many–one.

7 A possible path through the graph is:
10 10 10 10 9 10 10

$$\frac{10}{0} \to \frac{6}{0} \to \frac{6}{4} \to \frac{2}{4} \to \frac{2}{5} \to \frac{2}{5} \to \frac{7}{0} \to$$

0 4 0 4 4 3 3

9 9 9 9 9 9 9 10

$$\frac{7}{0} \to \frac{7}{4} \to \frac{3}{4} \to \frac{3}{5} \to \frac{8}{0} \to \frac{8}{3} \to \frac{4}{3} \to \frac{4}{3}$$

4 0 4 3 3 0 4 3

Chapter 3

❓ (Page 54)

There are several solutions, including:
Plymouth–Torbay–Exeter–Okehampton–Bude–Bodmin–Truro–Land's End
plus Exeter–Taunton–Minehead
plus Okehampton–Barnstaple
All solutions require 289 miles of cable.

Exercise 3A (Page 60)

1 MST has length 187.

2 MST has length 286. (There are two different MSTs.)

3 MST has length 463. (There are two different MSTs.)

4 MST has length 523.

Exercise 3B (Page 64)

1 MST has length 82.

2 MST has length 66.

Investigations (Page 65)

4 Using the bridge to link Hull to Barton, rather than linking Hull with Barton via Goole and Scunthorpe would save 22 km of cable.

6 (i) The most obvious way to link the towns is via crossroads converging at the centre of the square. This requires $4 \times 5 \sqrt{2} \approx 28.3$ km of road. However, the optimal solution is to introduce two points (known as Steiner points). If the square is defined by $(0, 0)$, $(1, 0)$, $(1, 1)$ and $(0, 1)$, then the Steiner points are at $(\frac{1}{2\sqrt{3}}, \frac{1}{2})$ and $(1 - \frac{1}{2\sqrt{3}}, \frac{1}{2})$, or a similar pair given by reflecting in $y = x$.

(ii) Similarly, an obvious way to link the chambers is by crossroads in three dimensions, converging on the centre of the cube but a more economical solution could be obtained using the principle of Steiner points in three dimensions.

❷ (Page 69)

The program has access to a database containing a list of locations (vertices), together with a list of distances (or times) between those locations that are linked directly (i.e. not via another location). These are the edges of the network. An efficient algorithm is then used to find the shortest (or fastest) route between any two specified locations.

Exercise 3C (Page 70)

1 (i) FPOMNJA; 24

(ii) BIHG; 25

(iii) EDCQA; 23

(iv) HIAQ; 22

(v) NKLQDC; 22

2 (i) CQLMOG; 21

(ii) GPFEDC; 26

3 FPOMNKL; 23
GOMNKL; 21
HIAJKL; 25

4 Before: GOMNKL; 21. After: GOMNJKL; 27. 6 mins longer.

5 OMNHI; 21

6 JKLMN; 15. Without KL – JAQLMN; 25. If delay > 10 minutes go round.

7 You could find the sum of the shortest times from C to each of the other points, and compare it with the sum of the shortest times from F to each of the other

points. C has the better mean time but this might not be the best criterion in practice. It might, for instance, be more important to be able to get to the centre of town more quickly.

Exercise 3D (Page 75)

1 (i) SPQRT or SUVWT; 15.

(ii) SABFEDT; 8.

(iii) SBFJT; 12.

2 (i) L.A.–San Fran–Salt Lake–Omaha–Chicago; 42.

(ii) New Orleans–Chicago–Omaha–Denver; 34.

(iii) L.A–Santa Fe–Denver–Omaha–Chicago; 42 (no better).
New Orleans–El Paso–Santa Fe–Denver; 31 (better).

3 C: $3 + 4 + 2 + 2 + 3 = 14$.

4 (i) (a) On to M5 at junction 6 and off at junction 10; distance 40.

(b) Route as above; time 49.5 minutes.

(c) 67.5 mph if joining at junction 5.

(d) Yes; join M5 at junction 7.

(ii) (a) On to M5 at junction 9; off at junction 6.

(b) Do not use motorway.

(c) Still use motorway, but on at junction 9 and off at junction 7.

Investigations (Page 79)

3 (i) 11 tons (SAET or SADT)
Algorithm: Permanently label that vertex which has largest temporary label. Update by using the rule 'temporary label = max (temporary label, min (new permanent label, weight))'.

(ii) SAFGT
Algorithm: Permanently label that vertex which has the smallest temporary label. Update by using the rule 'temporary label = min (temporary label, max (new permanent label, weight))'.

4 (i) Dijkstra fails because A is permanently labelled with 60 before the cheaper route of cost 50 is found. The cheapest route is SBACT (170). It may not be possible to overcome this problem. (For example, if the weight of AB were to be less than 20 then there would be no least cost path. The problem would be unbounded.) However, when all edges are directed, and when the problem is well-defined, a valid method is to assign a permanent label to a vertex only when all paths into that vertex are from permanently labelled vertices.

(ii) SEDT (2)

Exercise 3E (Page 84)

1 AB; BE; BD; BF; FC: 525.

2 (i) (2,4)–(6,1) (6,1)–(7,5) (7,5)–(11,9)
(7,5)–(11,3).
Total length ≈ 19.25.

(ii) (a) Min. connector is of length 6.
(b) Add in, say, (1,1). Gives a min. connector of
length $4\sqrt{2} \approx 5.66$.

3 (i) ACEJ 58 km.

(ii) Change distances to times.
Either add 5 to the weight of each edge incident on
E, so that any route through E has an increased
weight of 10; or delete E and all its edges and
compare the new minimum weight with 58 + 10.
ACGIJ 62 minutes.

4 (ii) e.g. CB; BM; DM; MA; AS; ES: total length = 300 m;
drains correctly, since there is no house at which
water accumulates.

(iii) Not worthwhile; can use CB; BA; AS; DE; ES with
total length = 270 m.

5 (i) (a) Update working values if min(label, edge
weight) > working value; label largest.
(c) GAFCB 9

(ii) Use DC. e.g. GDCB – 2

6 (i) BCDFGE cost = 18.

(ii) Best route is BAC with cost 2.
Dijkstra does not find it since C is labelled 3 at
stage 1.

7 (i) L–Am–B–A; length = 25.

(ii) (a) Maximise problem.
(b) Working values updated if greater.
Assign label when all arcs inward have been
considered.

8 (i) AB–6; ABC–12; ABCD–18; ABE–21; ABCF–18;
ABCG–16; ABCFH–21; ABCDI–24.

(ii) Increases distance and changes route to G;
ABCDG–23.

(iii) (a) Turn distances into times.

Add 10 minutes to every road incident upon C
(or do it twice, once with C and its arcs
deleted).

(b) ABEF–50 minutes.

9 (i) Mains–A; AC; CB; CD; CE length = 147.
(ii) Mains–E–D, length = 95.
(iii) (a) Mains–E–D.
Mains–A–C–B
Total length = 190.

(b) 43, 29.3%.

Chapter 4

❓ (Page 98)

1 The definition for total float gives the maximum
length of time slot during which the activity can be in
operation, minus the duration of the activity. Using it
may influence the flexibility that is available in
scheduling preceding or following activities.

The definition for independent float gives the
scheduling flexibility that is available without affecting
the flexibility that is available in scheduling any
preceding or following activity.

2 Earliest start time = earliest time for event i.
Earliest end time = earliest time for event i + duration.
Latest end time = latest time for event j.

Exercise 4A (Page 99)

1 36 hours: 1–2; 3–4; 4–5; 6–7; 8–9; 9–10.

2 170 days: A; B; F; J; R; U; V.
(i) 173 days: A; D; H; J; R; U; V.
(ii) I not critical, so no benefit.
(iii) Reduce B, since it is critical. New time is 165 days.
Same critical activities.
(iv) Time required is 177 days. Critical activities from
day 60 are K, R, U and V. Try to save two days by
speeding up one or two of these.

3

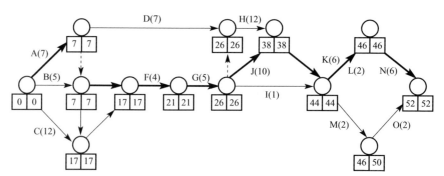

50 days: A; E; F; G; J; K; L; N.

4 18 weeks: B; E; F; H; J.

5 330s.

6 Arguably 16.5 minutes with B, D, F, I, K, L, M, O, Q, S
and U critical.

Exercise 4B (Page 107)

1 (i) Make and cook apple pie critical – 45 minutes.
Could do all tasks in 45 minutes with two people.
('Cooking' does not require resources. Making
batter and preparing potatoes must overlap with
making apple pie, otherwise activities can be
scheduled one at a time.)

(ii) Can certainly be done in 52 minutes with just one
person, but then it might not be sensible to have
the apple pie and custard ready at the same time as
the first course!

2 (i) 44 days: A; B; C; G; I; L; N.

(ii) D – 19 days (independent)
E – 20 days (independent)
F – 20 days (independent)
H/K – 1 day (shared)
J/M – 2 days (shared)
Sub-critical activities are H, J, K and M.

(iii) 6 workers are needed during days 24–33 inclusive.

(iv) With only five workers activity J is delayed by
12 days, until H has ended. J has a total float of 2,
so the total project time is lengthened from 44 to
54 days.

3 Duration is 18 days with critical path A, D and cost
£580.

(i) Reduce duration of A to 2 days at a total cost of
£680.

(ii) Reduce A to 6 days, D to 5 days and F to 2 days at
a total cost of £990.

Exercise 4C (Page 110)

1 (i), (ii) e.g.   A 1 to 2; B 1 to 3; C 3 to 4; D 5 to 6;
E 3 to 7; F 6 to 8; G 7 to 8; dummy 2 to 5;
dummy 4 to 5; dummy 4 to 7
whence early and late times are:
1 (0, 0); 2 (10, 10); 3 (3, 5); 4 (8, 10); 5 (10, 10);
6 (13, 13); 7 (8, 18); 8 (20, 20).

(iii) Min. completion time is 20 days.
Critical activities are A, D and F.

2 (ii) 9 days: B; E; G.

(iv) Start C at beginning of day 4 (i.e. at time 3).
Start F at beginning of day 5 (i.e. at time 4).

3 (i) Early time for event 6 = 15;
early time for event 8 = 19.

(ii) Late time for event 8 = 21;
late time for event 6 = 17.

(iii) Independent float for X = 0;   total float for X = 4.
Float represents scheduling flexibility.
Independent float is flexibility which can be used
without affecting the scheduling flexibility of
other activities.

(iv) Early time for event 52 = 19;
late time for event 50 = 17.
Do not know about other activities terminating at
event 58. Do not know about other activities
emanating from event 35.

4 (i), (ii)   (notation as in answer to question 1)
e.g. A 1 to 2;   B 1 to 3;   C 3 to 4;   D 3 to 5;
E 2 to 7;   F 5 to 7;   G 4 to 6;   H 7 to 8;
I 6 to 8;   dummy 2 to 3;   dummy 4 to 5;
dummy 6 to 7
whence: 1 (0, 0); 2 (6, 6); 3 (6, 6); 4 (12, 12); 5 (12,
16); 6 (18, 18); 7 (18, 18); 8 (22, 22).
Minimum time to completion = 22 days
Critical activities are A, C, G and H.

(iii) Latest start time for D is 13 (16 – 3). Earliest finish
time for C is 12.
So the project can be completed in 22 days.

5 (i) (a) 30 weeks; B; C; G.

(b) 40 weeks; A; F.

(c) Different critical activities.

(ii) (a) C can complete at 19.5.
E can complete at 19.
So G can start at 19, allowing the project to
finish at 31.5.

(b) Earliest start for C = 10.5.
Latest finish for C = 31.5.

6 (i)   A   B   C   D   E   F   G   H   I
      –   –   –   A,B   B   B,C   D   D,E,F   F

(ii) 9 days: B; D; G; H.

(v) 10 days: E and H.

7 (i) (b) 19 weeks: A; B; E; G; H; J.

(ii) (a) A1 A2 B1   B2   C1   C2   D1   D2
      – A1 A1 A2,B1 A1 A2,C1 A2,C1 C2,D1

(c) A2's float is 2 weeks. Total float for D1 and D2
is 6 weeks.

8 (i) (b) Min. duration = 90 minutes. Critical activities: A; B; G; H.

 (ii) (a) Make G dependent on E only, H dependent on B as well as F and G.

 (b) 75 minutes.

 (iii) e.g.

A	0	10	0	10
B	10	70	10	70
C	10	15	15	20
D	15	25	10	15
			20	25
E	25	45	25	45
F	45	55	45	55
G	55	70	55	70
H	70	75	70	75

9 (i) (b) Minimum duration = 11 days. Critical activities: B; C; E.

 (ii) 6 days.

 (iii) (b) 7 days.

 (iv) B has to be split into 2 activities, B_1 and B_2 with X dependent on B_1 and with C and D dependent on B_2.

10 (ii) Earliest start time for E is at $t = 0.5$ days. Latest start time for E is at $t = 6.25$ days.

 (iii) A; D; F; G; L. For critical activities late j time – early i time = duration.

 (iv) 7.5 days.

 (v) Add an activity Z of duration 4, dependent on F and preceding K. Critical activities are A, D, F, Z, K and L. Duration becomes 11 days.

11 (i) Needs H to be preceded by A, C, D, F and G. Needs K to be preceded by I and J. Needs L to be preceded by K.

 (iii) 63 minutes. D; H; J; K; L.

 (iv) 6.00 put pie in to cook
 6.00–6.05 peel potatoes
 6.05–6.07 wash mushrooms
 6.07–6.22 prepare salad
 6.10 put potatoes on to cook
 6.15 put mushrooms on to cook
 6.22–6.27 lay table
 6.30–6.45 dish up and eat first course
 6.45–6.50 eat dessert
 6.46– switch kettle on
 6.50–6.55 make and drink coffee
 6.55–7.03 wash up

 (v) According to the activity network, 5 minutes can be saved, with B/C becoming critical. In fact Martin has 27 minutes of work in preparing the first course, so only 3 minutes can be saved.

12 (ii) Minimum time to completion = 13 days. Critical activities: B; E; G; C; F.

 (iii) Crash E and C costing £1300: a 4.3% increase.

13 (ii) Minimum duration = 6 days. Critical activities: B; E.

 (iii) Either start C at $t = 2$, D at 3 and F at 4, or start D at 2, C at 3 and F at 4.

14 (ii) 53 minutes: B; C; F. Floats: A–5; D–7; E–7.

 (iii) B needs to be split into B1 and B2 with A dependent on B_1. 54 minutes: B1; A; C; F.

Chapter 5

❷ (Page 126)

The manager is constrained by the amounts of syrup, vitamin supplement and flavouring that are available. However, in this example there are many possible production plans that satisfy those constraints. The manager will need some criterion to decide which is the best. In this case the income generated from selling the drinks is to be maximised.

❷ (Page 130)

There are many others, for example (720, 0) and (0, 900), or any point on the line joining these two points.

❷ (Page 131)

Again, there are many possibilities. Any point on the line joining (728, 0) to (0, 910).

Exercise 5A (Page 133)

1 $x = 3\frac{1}{7}$; $y = 2\frac{6}{7}$; $P = 8\frac{6}{7}$

2 $x = 4\frac{1}{4}$; $y = 3\frac{1}{6}$; $P = 55\frac{1}{12}$

3 Let x be the number of minutes spent walking.
 Let y be the number of minutes spent running.
 Maximise $D = 90x + 240y$
 subject to $90x + 720y \leqslant 9000$
 $x + y \leqslant 30$
 Answer: $x = 20$; $y = 10$; $D = 4200$.

4 (i), (ii) Let a be the number of metres of cloth A produced.
 Let b be the number of metres of cloth B produced.

Maximise $P = 3a + 2.5b$

subject to $2a + b \leqslant 100$ (wool)

$\qquad 3a + 2b \leqslant 168$ (dye × 6)

$\qquad 5a + 4b \leqslant 360$ (loom time in minutes)

$\qquad 4a + 5b \leqslant 360$ (worker time in minutes)

(ii) Answer: $a = 17\frac{1}{7}$; $b = 58\frac{2}{7}$; $P = 197\frac{2}{7}$ (number of £ profit)

$41\frac{1}{7}$ minutes of loom time are unused. (Sufficient to show the constraint line not intersecting the feasible region.)

5 (i) Let a be the number of metres cut to plan A.

Let b be the number of metres cut to plan B.

Maximise $I = 11a + 12b + 8(200 - a - b) = 3a + 4b + 1600$ (or maximise $3a + 4b$)

subject to $0.05a + 0.07b \leqslant 12$

$\qquad a + b \leqslant 200$

(ii) Answer: $a = 100$; $b = 100$; $I = 2300$.

6 (i) Uses the fact that $x + y = 200$.

(ii) $36 \leqslant 0.45x + 0.15y \leqslant 72$

(iv) Cheapest: $x = 100$; $y = 100$.

Most expensive: $x = 140$; $y = 60$.

(v) $\frac{1}{10} \leqslant p \leqslant \frac{7}{10}$ giving $\frac{1}{2} \leqslant p \leqslant \frac{7}{10}$ as before.

Investigation (Page 135)

Lines of equal 'profit' are parallel to the boundary of the feasible region given by $5x + 4y = 100$.

If $5x + 4y = 100$, then $7.5x + 6y = 150$, so all solutions give an objective value of 150. Two vertices, $(8\frac{4}{7}, 14\frac{2}{7})$ and $(12, 10)$ give this objective value, so the assertion is not invalidated.

Exercise 5B (Page 137)

1 13 luxury and 17 standard, giving a profit of £335 000.

2 14 and 2 respectively, giving a profit of £6.20.

3 18 small cars and 10 large cars.

4 $x = 1$; $y = 2$; $z = 3$.

5 66 sprockets and 21 widgets.

6 (i) $\frac{200}{30} = 6\frac{2}{3}$ and $\frac{200}{40} = 5$.

$4x + 3y \leqslant 5100$

(ii) $4x + 5y \leqslant 5500$ (butter); $2x + 3y \leqslant 3000$ (sugar)

(iii) $I = 5x + 7y$

(iv) 750 biscuits and 500 buns, giving an income of £72.50.

Exercise 5C (Page 144)

1 Let w be the number of ounces of wheatgerm.

Let f be the number of ounces of oat flour.

Minimise $C = 8w + 5f$

subject to $2w + 3f \geqslant 7$

$\qquad 3w + 3f \geqslant 8$

$\qquad 0.5w + 0.25f \geqslant 1$

Answer: $w = 1.25$; $f = 1.5$; $C = 17.5$.

2 (i) $\quad 4x + 5y + s_1 = 45 \qquad 10x + 3y + s_1 = 52$

$\quad 4x + 11y + s_2 = 44 \qquad 2x + 3y + s_2 = 18$

$\quad x + y + s_3 = 6 \qquad\quad y + s_3 = 4$

(ii) (Only the two optimal vertices given here)

$\quad x = 3\frac{1}{7}$; $\quad y = 2\frac{6}{7}$; $\quad s_1 = 18\frac{1}{7}$; $\quad s_2 = 0$; $\quad s_3 = 0$

$\quad x = 4\frac{1}{4}$; $\quad y = 3\frac{1}{6}$; $\quad s_1 = 0$; $\quad s_2 = 0$; $\quad s_3 = \frac{5}{6}$

3 Let x be the number of tonnes of deep-mined and y the number of tonnes of opencast.

Maximise $\quad 10x + 15y \qquad$ (or minimise $10x + 5y$)

subject to $\quad x + y = 20\,000$

$\qquad\qquad 2x + y \leqslant 34\,000$ (chlorine)

$\qquad\qquad 3x + y \leqslant 40\,000$ (sulphur)

$\qquad\qquad 35x + 10y \leqslant 400\,000$ (ash)

$\qquad\qquad 5x + 12y \leqslant 200\,000$ (water)

Gives $x \approx 5700$; $y \approx 14\,300$, with water constraint critical and the others redundant.

4 Invest £20 000 in A and £30 000 in B, giving a return of 6.2%.

5 12 tables and 30 chairs; profit = £390.

It is likely that the demand for chairs will be greater than the supply.

6 Maximum = £140; minimum = £75.

7 (i) **(a)** Demand must be satisfied in month 1.

(b) Demand must be satisfied in month 2.

(c) No more than total demand may be produced in months 1 and 2.

(ii) Production plans: (5,4); (5,10); (9,0); (15,0).

(iii) Minimise $5p_1 + 6p_2$ (−12) from the total cost which is $p_1 + 5p_2 + 2(15 - p_1 - p_2) + 3(p_1 - 5) + 3(p_1 + p_2 - 9)$.

(iv) $p_1 = 9$; $p_2 = 0$; $(p_3 = 6)$

Cost = 33.

8 (i) Let x be the number of the cheaper jacket and y be the number of the more expensive jacket.

Maximise $P = 10x + 20y$

subject to $x + y \geqslant 200$

$\qquad 10x + 30y \leqslant 2700$

$\qquad 20x + 10y \leqslant 4000.$

(ii) $x = 186;\ y = 28;\ P = 2420.$

(iii) The purchase constraint exceeds the maximum order allowed by the other constraints $(220 > 214).$

9 (i) (a) $16x + 10y \leqslant 25.$

(b) $(0,0)$ feasible; $(0,1)$ feasible; $(1,1)$ infeasible; $(1,0)$ optimal.

(ii) (a) Eight possibilities, of which five are feasible.

(b) Optimum is $(0, 1, 1)$ with value £225.

(iii) (a) $2^{57} \approx 1.44 \times 10^{17}.$

Alternative answer: $^{57}C_{25}.$

(b) This is an integer programming problem. An LP does not usually give an integer solution.

10 (i) Minimise $265a + 205b$

subject to $a + b \geqslant 45$

$\qquad 13a + 10b \geqslant 500.$

(ii) $a = 16\frac{2}{3};\ b = 28\frac{1}{3};\ \text{cost} = £10\,225.$

(iii) $a = 10;\ b = 37.$

(iv) Big difference in worker mix. Not much difference in cost (£10).

11 (i) Let a hours be spent on A and b hours on B.

$a + b \leqslant 6$

$a \geqslant 2$

$b \geqslant 2$

$a = 1.5b.$

(ii) $a \geqslant 2$ is redundant.

Feasible points consist of the line segment joining $(3, 2)$ to $(3.6, 2.4).$

(iii) (a) $(3.6, 2.4)$

(b) $(3, 2)$

(iv) Putting $a = 1.5b$ in the inequalities gives $b \leqslant 2.4,$ $b \geqslant \frac{4}{3}$ and $b \geqslant 2,$ giving the same answer.

12 (i) The constraints to be graphed are:

$25x + 100y \geqslant 15\,000$

$12x + 60y \leqslant 12\,000$

$x \geqslant 200$

$y \geqslant 50$

(ii) $(25 - 12)x + (100 - 60)y.$

(iii) $x = 750;\ y = 50;\ £11\,750.$

(iv) $x = 400;\ y = 50;\ £7800$ and $£7200.$

Chapter 6

❓ (Page 150)

Mean = 3; most likely = 1.

Exercise 6A (Page 151)

The answers to the questions in this exercise depend on the results of your simulations.

1 (iii) You need to be able to generate a realisation of a uniformly distributed discrete random variable which can take values 1, 2, 3, ... , 10. Ideas based on throwing a die twice and adding the scores will not work since the outcomes will not be equally likely (uniformly distributed).

The following approach is based on place values. There are easier ways for sampling from 1, 2, 3, …, 10, but this method generalises easily. Throw a die, subtract 1 from the total and record the result as a numbers of units. Throw it again, subtract 1 from the total and record it as a numbers of 6s. This procedure gives a base 6 number between 0 and 55, which can be converted to a denary (base 10) number between 0 and 35. If the denary result is between 1 and 10 take that as the outcome. If it is between 11 and 20, subtract 10 and take that as the outcome. If between 21 and 30, subtract 20 and take that as the outcome. Otherwise ignore and repeat the process.

2 (iii) For example, the arrival pattern of pedestrians at a real crossing could be studied and the time intervals between arrivals recorded.

There could be a minimum time between the commencement of 5-second crossing periods, so that road traffic is not seriously disrupted.

3 (i) For example, the Excel cell formula = IF(RAND()<0.5, "H", "T") will place H or T in the cell, each with probability 0.5.

(iii) (a) e.g. = 1+INT(6*RAND())

(b) e.g. = INT(10*RAND())

6 In the absence of data, suitable assumptions might be as follows. A first assumption might be that patients arrive 0, 1, 2, 3, 4 or 5 minutes early, each with probability $\frac{1}{6}$, so that a die could be used to simulate arrival times. Consultation times might also, in the first instance, be assumed to be uniformly distributed.

It must be recognised that these assumptions are only made for the purpose of the exercise. In reality the distributions of arrival times and consultation times would have to be investigated.

Exercise 6B (Page 156)

The answers to the questions in this exercise depend on the results of your simulations.

3 You probably found that a queue was building up when there was only one petrol pump, but not when there were two. However, since you are dealing with randomness it is possible that you did not find this. Try running the simulation again with different random numbers.

Exercise 6C (Page 157)

The answers to the questions in this exercise depend on the results of your simulations.

1 (i) 6.3 minutes. Two car washes can deal with 2 cars every 12 minutes, or with a car every 6 minutes. They will just about be able to cope, though there will be some queueing.

3 (ii) Mean arrival interval = 1.79 minutes.
Mean service time = 2.78 minutes.
A second counter is needed. One server will not be able to cope.

5 (i) 0.167 0.233 0.2 0.067 0.1 0.067 0.067 0.033 0.067 0
0.167 0.4 0.6 0.667 0.767 0.833 0.9 0.933 1 1

(ii) Using (approximate) interval mid-points:
30s; 30s; 50s; 10s.

(iii)
Customer	1	2	3	4
Arrives (secs)	30	60	110	120
Service start	30	60	110	
Service end	60	80	138	

Queue empty from time 80 to time 110 i.e. for 30 seconds.

6 (iii) For example, order $x - n$ televisions, where n is the number in stock at the end of the week and x is the number sold during the week.

7 (iii) The most important assumption is the distribution of demand. This needs to be investigated, but for the purpose of the exercise you probably assumed it to be uniformly distributed between 1500 and 2500 gallons per day. You probably also assumed that there was no order pending at the beginning of the simulated period.

Exercise 6D (Page 162)

1 (i) (00–24) →2; (24–74) →3; (75–99) →4
3; 2; 2; 4; 2

(ii) (00–31) →1; (32–79) →2; (80–95) →5; (96–99) ignore
2; 5; 2; 1; 1; 1

(iii)
Customer	1	2	3	4	5	6
Arrives (mins)	0	3	5	7	11	13
Service start	0	3	8	10	11	13
Service end	2	8	10	11	12	14

(iv) Mean time in queue = 1 min
$$\text{Mean length of queue} = \frac{(0\times5)+(1\times2)+(2\times1)+(1\times2)+(0\times4)}{14}$$
$$= \tfrac{6}{14} \approx 0.43 \text{ people}$$

(or $\tfrac{6}{14}$ directly by using 6 person minutes spent queuing).
Server utilisation = $\tfrac{12}{14} \times 100 \approx 86\%$

(v) Extended run.

2 (i) e.g. (00–27) → wet; (28–97) → dry; 98, 99 ignore.

(ii) Using above: D; D; D; W; W; D; W; W; D; D; W; D; D; D.

(iii) $\tfrac{5}{14}$.

(iv) $\tfrac{7}{14}$ (or $\tfrac{8}{14}$ if initial transition is included).

(v) Better model (not just wet/dry); more repetitions.

3 (i) Large proportion (94%) of realisations rejected.

(ii) For example, (00–47) →2; (48–63) →7; (64–95) →35.

(iii) 2; 2; 7; 7; 7; 35; 2; 2; 35; 35; 2; 2; 35; 2; 35

(iv) Cumulative times: 2; 4; 11; 18; 25; 60; 62; 64; 99.
1	2	3	4	5	6
0	10	20	30	40	50
25	29	33	37	41	50

(v) Max. queue size = 3 (from time 20 to time 25).

(vi) More detailed distribution for intervals between cars on main road; better model for interval between cars on side road; more repetitions.

4 (i) (00–09) →1; (10–49) →2; (50–69) →3;
(70–89) →4; (90–99) →5

(ii)
G1 times:	4	2	1	2	3	2
G2 times:	5	1	3	2	4	4
Totals:	9	3	4	4	7	6

Mean waiting time = 5.5

(iii) (a)
Totals:	7	10	10	12	5	9

(b)
Mins:	7	3	4	4	5	6

Mean time = 4.83.

(iv) Accounting for group size; considering other starting conditions; shorter time slices, e.g. 30 seconds; more repetitions.

5 (i) **(a)** (00–09); (10–49); (50–69); (70–89); (90–99)

     (b) 2.0; 2.5; 1.0

  (ii) **(a)** (00–31); (32–79); (80–95)

     (b) 0.5; 0.5; 1.0

(iii) Rule 1

2	12:00	12:02	12:02	12:02.5
1	12:01.5	12:04	12.04	12:04.5
2	12:04	12:05	12:05	12:06

Rule 2

2	12:00	12:02	12:02	12:02.5
2	12:02	12:04.5	12:04.5	12:05
1	12:02	12:03	12:04.5	12:05

  (iv) Service of all three cars completed earlier ... at expense of car 2.

Index